Vacuum Technology

Second Edition

Vacuum Technology

An introduction

L G Carpenter MA, BSc, FInstP

Second Edition

Adam Hilger Ltd, Bristol

British Library Cataloguing in Publication Data

Carpenter L G
    Vacuum Technology.—2nd ed.
    1. Vacuum technology
    I. Title
    621.5'5   TJ940

    ISBN 0-85274-481-1

Published by Adam Hilger Ltd,
Techno House, Redcliffe Way, Bristol BS1 6NX

The Adam Hilger book-publishing imprint is owned by The Institute of Physics

Printed in Great Britain by
Page Bros (Norwich) Ltd

Preface to the First Edition

The aim of this book is twofold: first, to provide for those who are about to use vacuum technology as a tool for the first time an introduction to its principles, before they consult more specialised works and the literature, and second, to present a broad view of the field to those who for professional reasons need a general knowledge of the subject.

This double aim has imposed omission of detail, brevity of style and, in some cases, over-generalisation, the evils of which I hope may be mitigated by the references given to more specialised books and papers.

Appendices I and II are intended for the convenience of those readers who may wish to be reminded of the physical formulae and terms used in the text.

I gratefully acknowledge my indebtedness to many friends and colleagues, including Mr G I Lewis, Mr W N Mair, Mr B D Power, Mr T W G Rowe, Mr P Ridgway Watt and particularly Dr G W Green, who read the whole of the first draft and made many constructive criticisms and suggestions. Mr M J Watts kindly read the proofs, but responsibility for remaining errors is entirely mine.

My thanks are due to Professor J P Roberts, whose invitation to lecture on vacuum technology in the Department of Ceramics at the University of Leeds led to this book, and to Professor D W Holder for stimulating discussions and for the hospitality of his laboratory.

To the publishers, and especially to Mr Neville Goodman, whose helpful patience has been unfailing, I am deeply grateful.

L G Carpenter

Department of Engineering Science
Oxford

June 1969

Preface to the Second Edition

The aim of this edition is the same as that of the first, namely to provide a brief introduction to vacuum technology and to supply literature references so that topics of special interest to the reader can be pursued in depth.

The framework of the book is unchanged, but additions, corrections and updatings have been made.

L G Carpenter

September 1982

Contents

1 Vacua and Vacuum Technology

1.1 What is a Vacuum?

The term vacuum is generally used to denote a portion of space in which the pressure is significantly less than 1 bar†. The definition has been refined by the British Standards Institution as follows:

Coarse or rough vacuum	760–1 Torr
Medium vacuum	$1-10^{-3}$ Torr
High vacuum	$10^{-3}-10^{-6}$ Torr
Ultra-high vacuum	10^{-8} Torr and below.

From the physical point of view, a high vacuum exists where the mean free path (see Appendix I, § AI.9) of the molecules, as calculated from the pressure, is much greater than the linear dimensions of the vessel. The situation is then dominated by gas–wall (as opposed to gas–gas) collisions. Viscosity, as ordinarily defined, has little reference or meaning in these circumstances, the basic concept being one of random wandering from one gas–wall collision to another without 'memory' of the incident direction before collision. However, one aspect of vacuum technology which does not fall within this definition of a vacuum is the action of vapour-stream (diffusion) pumps whose action depends on gas–gas collisions. This is dealt with in § 3.3.

Formerly the torr was in common use as a unit of pressure. However, it is now (1982) more usual to employ the pascal or the millibar, the relation between these units being:

$$1 \text{ mbar} = 0.75 \text{ Torr}$$
$$1 \text{ Pa} = 7.5 \times 10^{-3} \text{ Torr.}$$

---

† 1 bar = 750 Torr, the torr being approximately the pressure due to a column of mercury 1 mm in height.

1

1.2 The Basic Topics of Vacuum Technology

The creation of a vacuum involves two processes, the first being the removal of the gas originally contained in the volume of the work chamber, and the second a competition between the capacity of the pump and the production of gas, not originally in the gas phase, but stemming from leaks—real or virtual—the latter being gas desorbed (see Appendix II, § AII.4) from the walls and contents. Hence the topics to be treated include the speed of pumps and the conductance of channels by which they are connected to the work chamber (Chapter 2); their construction and mode of operation (Chapter 3); means of determining the pressures achieved and the composition of the gases composing them (Chapter 4); the design and choice of materials to minimise leaks, both real and virtual (Chapter 5); and the rather special techniques needed for producing and measuring the highest vacua (Chapter 6). Some practical applications of vacuum technology are described in Chapter 7.

2 Conductance and Pump Speed

2.1 The Concept of Conductance

Conductance (F) is concerned with the speed of gas flow through an aperture or channel under unit pressure difference. It is somewhat analogous to electrical conductance in that, as will become evident in § 2.2, the reciprocals of conductances in series (i.e. resistances) are additive. The definition is as follows:

$$F = \frac{\text{flux of gas molecules in } PV \text{ units† per second}}{\text{pressure difference across the aperture or channel}}.$$

Thus, if V_1 is the volume per second‡ (measured at the pressure P_1) traversing the aperture or channel, across which a pressure difference $P_1 - P_2$ exists,

$$F = \frac{P_1 \dot{V}_1}{P_1 - P_2}. \tag{2.1}$$

For a perfect gas (see Appendix I, equation (AI.8))

$$P_1 \dot{V}_1 = n \dot{V}_1 k T \qquad P_1 - P_2 = (n_1 - n_2)kT \tag{2.2}$$

where n denotes the number of molecules per cm^3, k is Boltzmann's constant and T is the absolute temperature. Then

$$F = \frac{n_1 \dot{V}_1}{n_1 - n_2} \tag{2.3}$$

$$= \frac{\text{flux of molecules per second}}{\text{difference in molecular density}} \tag{2.4}$$

and has dimensions $[\text{Length}]^3 [\text{Time}]^{-1}$, i.e. volume per unit time.

† For meaning of PV units, see Appendix I, § AI.7.
‡ The notation $\dot{x}$ is used throughout for dx/dt.

2.2 Conductances in Series and Parallel

It is sometimes convenient in design calculations of gas flow through channels to work with the reciprocal of F, namely the impedance (W) and to write

$$\dot{Q} = P/W \qquad (2.5)$$

where $\dot{Q}$ is the gas flow in PV units per unit time and P the pressure difference across the channel.

Equation (2.5) is analogous to Ohm's law, $\dot{Q}$, P and W being analogies of current, voltage and resistance in electrical circuits.

For an assembly of channels *in series* the resultant impedance (W_0) is

$$W_0 = \sum W_i \qquad (2.6)$$

where W_i is the impedance of the ith channel.

For a number of channels *in parallel* the total flow is clearly equal to the sum of the individual flows, and

$$\dot{Q} = \sum P/W_i \qquad (2.7)$$

which, since the pressure is the same across each channel, becomes

$$\dot{Q} = P \sum 1/W_i. \qquad (2.8)$$

However, if W_0 is the resultant effective impedance of the whole assembly

$$\dot{Q} = P/W_0. \qquad (2.9)$$

Hence, comparing (2.8) and (2.9)

$$1/W_0 = \sum 1/W_i. \qquad (2.10)$$

The practical consequence of (2.6) is that the overall conductance of a number of conductances in series is dominated by the smallest conductance (largest impedance) of the assembly, in analogy with the dominance of the largest resistance in a chain of series resistances in the electrical case.

2.3 The Effect of Pressure on Flow Regime and Conductance

Three regimes of gas flow may be distinguished. At high pressure— where λ (the mean free path) is much less than d (a typical dimension of the vessel)—there is Poiseuille flow, which is dominated by viscosity, i.e. by inter-molecular collisions. At the other extreme ($\lambda \gg d$) there is free molecular flow, in which gas-molecule–wall collisions are dominant. The intermediate case is called transition flow, where both mechanisms are operative.

For molecular flow, F is independent of pressure, and dependent only on geometry and on the molecular weight and temperature of the gas; it is therefore a simple and most useful concept. Molecular flow alone will be dealt with here, because it is truly a vacuum phenomenon (in the sense of § 1.1), and because it is the conductance in the low-pressure molecular flow channels which, except in very large chambers or in pumping large continuous gas loads, usually determines the efficiency of the pumping system as a whole.

Conductance in the Poiseuille and transition regimes is a function of pressure as well as geometry and temperature. Values of conductance in the three regimes are given for example by Dennis and Heppell (1968) and by Lewin (1965, pp 14 ff).

2.4 The Conductance of an Aperture

Consider the case of molecular flow through an aperture of area A in a diaphragm of area large compared with A, separating two vessels at the same temperature in which the molecular densities are n_1 and n_2 molecules/cm^3 respectively, and the mean speeds (being determined by the temperature) have a common value $\bar{c}$ cm s^{-1}. From equation (AI.6), the numbers of molecules which 'hit' the aperture from the two sides[†] and pass through it are $\frac{1}{4}An_1\bar{c}$ and $\frac{1}{4}An_2\bar{c}$ respectively. Hence the net flux is $\frac{1}{4}A\bar{c}(n_1 - n_2)$ and from equation (2.4)

$$F = \frac{A\bar{c}}{4}\frac{(n_1 - n_2)}{(n_1 - n_2)}$$

$$= \tfrac{1}{4}A\bar{c}. \tag{2.11}$$

Note that F is directly proportional to area and independent of shape.

---

† Assuming the Maxwellian speed distribution undisturbed by the flow.

The precise validity of the formula depends upon the edges of the aperture being infinitely thin, otherwise some of those molecules which hit the edge of the aperture fail to pass through it. The magnitude of this failure depends on the ratio of the thickness of the aperture walls to its linear dimensions.

At 25°C, $\bar{c}$ for N_2 is 4.745×10^4 cm s^{-1}, hence F for an aperture of 1 cm^2 is about 11.9 l s^{-1}. Equation (AI.3) shows that $\bar{c}$, and therefore F, is proportional to $(T/m)^{1/2}$ where m is the mass of the molecule.

2.5   The Conductance of a Channel

For molecular flow through a long channel, Dushman (1962, p 88) gives the conductance as

$$F = \frac{4}{3} \bar{c} \frac{1}{\int_0^l H/A^2 \, dx} \tag{2.12}$$

where H is the perimeter, A the cross section area at any point x, and l is the total length.

For the particular (and common) case of a long tube of constant radius a, formula (2.12) becomes, if a and l are in cm,

$$F = \frac{2}{3} \frac{\pi a^3 \bar{c}}{l} = 30.48 \frac{a^3}{l} \left(\frac{T}{M}\right)^{1/2} l s^{-1} \tag{2.13}$$

where T is the absolute temperature in Kelvins and M is the molecular weight. A useful approximation for air at 25°C is

$$F = 10^2 \, a^3/l \, l s^{-1}. \tag{2.14}$$

Equation (2.13) gives the conductance of a long tube, and ignores the fact that, even if the tube were infinitely short (i.e. merely an aperture), it would still have (see equation (2.11)) a conductance of $\frac{1}{4}\pi a^2 \bar{c}$. Remembering that the reciprocals of conductances are additive it may be shown that the total conductance is approximately obtained by adding a length $8a/3$ to the geometric length l of the tube. Thus for short tubes (a comparable with l), equation (2.13) overestimates the conductance. The end effect is shown by the curvature in the large diameter, small length region of figure 2.1, which gives curves for the conductance of circular tubes for air at 20°C. Right angle bends do not materially decrease the rate of molecular flow (see Lewin 1965, p 16).

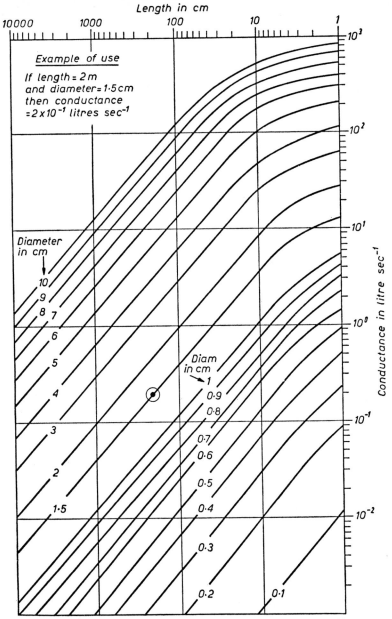

Figure 2.1 Conductance of cylindrical tubes for air at 20°C.

The mechanism of conductance is discussed from the molecular point of view in Pollard and Present (1948), and Steckelmacher (1966) has given an excellent critical review of work on molecular flow conductance for systems of tubes and components.

2.6 The Conductance of a Cold Trap

It is often desirable to interpose a tortuous cooled channel between a diffusion pump (see § 3.3) and the work chamber being pumped. Its purpose is to condense the working fluid of the pump at a temperature such that its saturated vapour pressure at this temperature is acceptably low in the work chamber. Such a channel is called a 'cold trap'.

To be effective, a cold trap must be at a sufficiently low temperature and of such geometry that a vapour molecule of pump fluid cannot get from the pump to the work chamber without hitting the wall of the trap and condensing upon it for a time that is long compared with the duration of use of the trap. Equally, a molecule of the gas to be pumped cannot traverse the trap from the work chamber to the pump without hitting the wall of the trap and being cooled thereby, i.e. its speed will be reduced. Thus both opacity and reduction of velocity impede the pumping of gas molecules, so in calculating the total conductance of the pumping system the conductance of the cold trap must be taken into account. The conductance has been computed in some simple cases (see e.g. Dushman 1962, p 101). In applying formulae valid for air at room temperature, allowance must be made for the effect on $\bar{c}$ of the trap temperature and also of the molecular weight of the gas, if it is other than air.

The conductance of traps may be approximately predicted by an optical analogue method (Carpenter and Watts 1972, 1976, 1982). This exploits the similarity between the scattering of molecules from a rough surface and the scattering of light from a diffusely reflecting surface.

2.7 The Speed of a Pump

A pump is a device with an inlet port having the property that a certain fraction of the molecules of the gas to be pumped which enter it do not return. Its speed (S) may be defined as follows:

$$S = \frac{\text{net flow per second entering the pump port, measured in } PV \text{units}}{\text{pressure } (P) \text{ at the entry port}}$$

$$= \frac{P \times \text{volume per second, measured at pressure } P}{P}$$

= net volume per second (measured at the entry pressure) traversing the entry port. (2.15)

The precise definition of P is discussed in § 3.34. If A is the area of the entry port, and v is the average velocity measured perpendicular to A

$$Av = S \qquad (2.16)$$

or

$$v = S/A$$
= pump speed per unit area of entry port. (2.17)

If every molecule which crossed the plane of the entry port and entered the pump failed to return, the maximum possible value of v would be $\bar{c}/4$ (equation (AI.6)) so that the speed of a pump for N_2 at 25°C could not exceed $\frac{1}{4} \times 4.75 \times 10^4 \, \text{cm}^3 \text{s}^{-1}$ per cm^2 area of entry port, i.e. $11.9 \, \text{ls}^{-1} \text{cm}^{-2}$.

If, as is frequently the case, there is a channel of conductance F between the entry port of the pump and the work chamber, and if the flow is in the molecular regime, so that the flux through F is linear in the pressure differential across it, then

$$PS = (P_c - P) F = \text{flux in } PV \text{ units}$$
$$= S_c P_c$$

where

P = pressure at entry port of pump
S = speed of pump
P_c = pressure in work chamber
S_c = effective pump speed *at the work chamber*.

Algebraic elimination of the pressures from the above equations leads to

$$S_c = S \frac{F}{F + S}. \qquad (2.18)$$

Hence, the effective speed S_c is always less than S, the speed of the pump itself, and this reduction can be serious; e.g. if $F = S$, the introduction of F between the pump and the work chamber will halve the effective pumping speed†.

If $F \ll S$ equation (2.18) becomes

$$S_c \simeq F. \tag{2.19}$$

A very common design fault in vacuum systems is the incorporation of a large, and therefore, expensive pump connected to the work chamber to be evacuated by a channel of a conductance so small that much of the potential speed of the pump is wasted.

2.8   Pump-down Time

Consider a work chamber of volume C, which has a constant leak L (measured in PV units per unit time), evacuated by a pump having an effective speed S_c at the chamber. After an infinite time a limiting pressure P_∞ will be produced in the chamber. At any time t and pressure P the rate of extraction of gas from the volume C, plus the rate of entry via the leak, must equal the overall rate of extraction by the pump, if there is no other source of gas. Thus

$$-C\dot{P} + L = S_c P. \tag{2.20}$$

The solution of (2.20) is

$$2.3 \lg \left(\frac{P - P_\infty}{P_0 - P_\infty} \right) = - \frac{S_c t}{C} \tag{2.21}$$

where P_0 is the initial pressure, and P is the pressure at any time t after the start of pump-down, and $P_\infty = L/S_c$.

Thus P decays exponentially to P_∞ with a time constant of C/S_c.

As an example take

$$S_c = 10 \, \text{l s}^{-1}$$

$$C = 50 \, \text{l}$$

$$P_\infty = 10^{-1} \, \text{mbar}$$

---

† S, the speed of a pump, is a quantity of the same nature as F, the conductance of a channel. In the former, $S \times P$ (which may be written as $S(P - 0)$), is the net flux into the pump, in the latter, $F(P_1 - P_2)$ is the net flux through the channel.

corresponding to $L = 1 \text{ mbar l s}^{-1}$ (a large leak!)

$$P_0 = 10^3 \text{ mbar}$$

$$P = 1 \text{ mbar}.$$

Equation (2.21) then predicts the time t to pump-down to 1 mbar as 35 s.

Note that S_c is the effective speed at the work chamber and is therefore in principle (see equation (2.18)) a function of F (the conductance between work chamber and pump) as well as of S, the speed of the pump itself. However, in the Poiseuille and transition flow regimes, often $F > S$ so that, in practice, $S_c \approx S$ and is virtually constant. If this is not so, or if S itself varies with P over the range of interest, the total pump-down time may be obtained by summing the times over successive pressure ranges, using the appropriate values of F and S to obtain S_c. The usefulness of equation (2.21) is limited to pressures above which out-gassing or virtual leak (see § AII.4) as opposed to the 'real' leak L † make significant contributions to the gas load on the pump. In practice this means pressure above (say) $10^{-1}–10^{-2}$ mbar, the exact value depending on the 'gassiness' of the walls of the chamber and its contents.

References

Carpenter L G and Watts M J 1972 *J. Phys. D: Appl. Phys.* **5** 1374
——1976 *J. Phys. D: Appl. Phys.* **9** 26–39
——1982 *Vacuum* **32** 307–8
Dennis N T M and Heppell T A 1968 *Vacuum System Design* (London: Chapman and Hall) pp 212 ff
Dushman S 1962 *Scientific Foundations of Vacuum Technique* 2nd edn, ed J M Lafferty (New York: Wiley)
Lewin G 1965 *Fundamentals of Vacuum Science and Technology* (New York: McGraw-Hill)
Pollard W G and Present R D 1948 *Phys. Rev.* **73** 763
Steckelmacher W 1966 *Vacuum* **16** 561

---

† The justification for assuming, in equation (2.20), that L is independent of P is that for large P it is small compared with $C\dot{P}$ and therefore without influence, while for small P (where it is important) it becomes pressure independent.

3 Vacuum Pumps

3.1 Preliminary Survey of Types of Vacuum Pump

According to their mode of operation, pumps may be classified as:

(*a*) *Mechanical pumps* in which gas is trapped, compressed and removed bodily from the low-pressure to the high-pressure side of the pump, whence it is expelled to the atmosphere either directly or through a second mechanical 'backing' pump.

(*b*) *Mechanical pumps* which impart velocity to the pumped gas molecules by impact with swiftly moving solid surfaces. Such pumps are usually backed by another mechanical pump interposed between them and the atmosphere.

(*c*) *Vapour-stream pumps*† in which the gas molecules to be pumped are caused to move in the desired direction by impact with heavy molecules of pump fluid vapour, which derive their velocity by boiling from a liquid reservoir. With the exception of steam ejector pumps (which exhaust directly to the atmosphere) vapour-stream pumps are backed by mechanical pumps.

(*d*) *Chemical pumps* in which the molecules to be removed from the work chamber are caused to combine chemically with highly reactive substances ('getters'), such as titanium, thus being converted to and trapped in the solid phase. The getters are generally dispersed on the pump surfaces by evaporation or by sputtering.

(*e*) *Sorption pumps* in which the molecules to be pumped are physically sorbed (e.g. by van der Waals forces) on specially prepared substances of large specific area (e.g. charcoal or zeolites). The sorption is aided by cooling the sorbent below room temperature and is reversible with temperature.

(*f*) *Cryopumps*, which are vessels whose internal chemically non-reactive metal surfaces are cooled by refrigerators to a temperature such that the gases to be pumped condense upon them. After the condensation of one or two monolayers, the process is indistinguishable from the condensation of a substance on a substrate of itself, thus differentiating it from (*e*) above.

---

† Often called diffusion pumps.

In the pump types (*a*), (*b*) and (*c*), the pumped gas is eliminated from the system; in the other types it is retained.

3.2 Mechanical Pumps

3.2.1 Oil-sealed mechanical pumps

These pumps, which are very widely used, belong to category (*a*) above. They run immersed in oil, which lubricates and seals the rubbing interfaces and reduces the dead spaces at the end of the compression, thus improving the compression ratio.

The rotating vane type, common for smaller pumps, is illustrated schematically in figure 3.1. Gas entering the pump from the work chamber is trapped and compressed by the rotation of rotor 3 in stator 4. Gas tightness of the rotor/stator interfaces is maintained by the vanes 1 and 2 (forced outwards by the spring 5 and by centrifugal force) and by the small clearance between the two ends of the rotor and the stator, the seal being improved by the small quantity of oil on the stator and rotor surfaces.

The trapped gas is conveyed to the compression side of the pump and is expelled, together with a little oil, when the pressure exceeds atmospheric pressure sufficiently to raise the exit flap valve 6.

If the gas to be pumped contains a condensable vapour, contamination of the oil and reduction of the vacuum attainable result. As a specific example, consider the pumping of air which contains some water vapour. As the inlet pressure to the pump decreases during the pump-down, the compression ratio necessary to raise the pressure of the trapped gas to a value sufficient to open the exit valve against the combined resistance of its spring and the atmosphere increases. If the compression ratio is such that the partial pressure of the water vapour is raised to a value equal to the saturated vapour pressure of water at the temperature of the pump, the vapour will condense and may not be completely expelled or will be expelled, via the exit valve, as liquid into the oil. Thence it will find its way back into the body of the pump (via the oilways provided for maintaining the lubrication and the oil seal) and will re-evaporate, thus setting a limit to the vapour pressure of water in the pumped chamber. The device of gas ballast reduces this unwanted effect by introducing atmospheric air at a point in the cycle (such as that shown by the rotor position in figure 3.1) where the gas being pumped is isolated from the inlet port, but is connected to the spring-loaded valve 7. This valve admits air, the compression of which, by further rotor movement, is sufficient to lift the exit valve 6 and eject ballast air, gas to be pumped and water vapour while the latter is still in the vapour phase and can pass right out of the pump. The amount of air ballast is adjustable by valve 8. An account of the theory and practice of gas ballasting is given by Power (1966, pp 21 ff), whose book should be consulted for an authoritative account of oil-sealed rotary pumps and, indeed, pumps of all types.

A typical single-stage rotary pump pumping air might have an ultimate pressure of 5×10^{-3} mbar, as measured by a gauge sensitive only to the

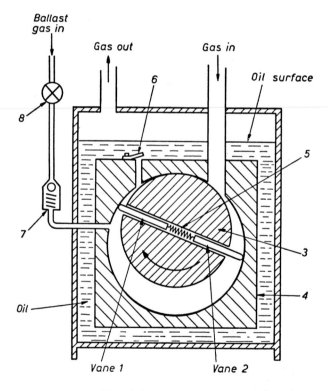

Figure 3.1   Oil-sealed rotary vane pump, gas ballasted.
1, vane; 2, vane; 3, rotor; 4, stator; 5, spring; 6, flap valve; 7, spring-loaded valve; 8, valve.

permanent (non-condensable) gases, but nearly an order of magnitude higher if the gauge is sensitive to the vapour pressure of the oil and its decomposition products. Its speed would start to decrease significantly at about 1 mbar. For this reason, two rotary pumps (constituting a two-stage pump) are often put in series and mounted as one unit. Typically, the speed of a small unit might be $1\,l\,s^{-1}$ when driven at 650 rpm by a $\frac{1}{3}$ hp motor, decreasing significantly below about 10^{-2} mbar and becoming zero at its ultimate pressure (for permanent gases) of 10^{-3} mbar. With gas ballast, speed decrease sets in at a higher pressure, and the ultimate pressure is increased by (typically) a decade. The range of sizes available corresponds to a range of speed of about $1{-}100\,l\,s^{-1}$.

3.2.2 Roots pumps

These pumps, also called mechanical booster pumps, belong, as is evident from figure 3.2, to category (*a*) of § 3.1, since gas is bodily removed by the contra-rotating figure-of-eight-shaped rotors from the low-pressure to the high-pressure side. A mechanical booster pump is generally backed by another pump interposed between it and the atmosphere. Hence the pressure difference across it is much less than 1 bar, and oil (with its usually undesirable vapour pressure) is not

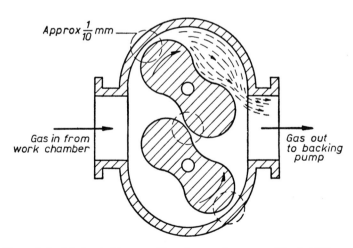

Figure 3.2 Mechanical booster (Roots) pump (courtesy of Leybold Heraeus Ltd).

necessary for sealing the clearance ($\sim 10^{-1}$ mm) between rotors and stator, but only for lubricating the gears, which ensure the necessary very accurate synchronisation of the rotors. These pumps are suitable for high-speed pumping in the pressure range 10^{-2}–10 mbar. Their characteristics and use are discussed by Power (1966, ch 5). Speeds range from about 50 l s^{-1} to more than 100 times this value.

3.2.3 Mechanical molecular pumps

These pumps fall into category (*b*) of § 3.2, and depend on the impingement of molecules on a moving surface and their re-emission, having acquired some of the macroscopic velocity of the moving surface superimposed on their Maxwellian velocities. Suitable ducting arrangements direct the flow of molecules from the low-pressure to the high-pressure side of the pump. Two examples are given below.

3.2.3.1 Molecular drag pumps. These were introduced in 1913 by W Gaede, the principle of whose pump is shown schematically in figure 3.3. Gas entering port P is expelled at port E, having acquired an average velocity in a clockwise direction by repeated impact with the surface of rotor A, moving at about 5000 cm s^{-1}. The subsequent development of rotary drag pumps by Holweck, Siegbahn and Gondet is described by Power (1966, pp 190 ff) who points out that their most

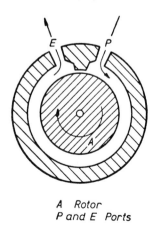

A Rotor
P and E Ports

Figure 3.3 Principle of Gaede's molecular pump.

suitable applications exploit either their immediate availability (without warm-up period) and comparatively vapour-free vacuum or their characteristic of pumping heavy molecules better than light ones†. A disadvantage of these pumps is the necessarily small clearance between rotor and stator. The turbo-molecular pumps described below have clearances of several millimetres.

3.2.3.2   Turbo-molecular pumps. A typical pump (see Becker 1966, pp 625 ff) consists (figure 3.4) of a metal housing in which a rotor runs at 16 000 rpm, supported on water-cooled oil-lubricated bearings at each end of the shaft. Obliquely slotted discs are arranged alternately on the rotor and stator. These are analogous to the rotor and stator blades of a turbine but are simpler in that, at low pressures, the exact aerodynamic shape is not significant. The rotation of the rotor blades urges the gas from the central entry port to each end of the pump whence it flows to the exit port, carrying with it any oil molecules from the lubricated bearings which would otherwise backstream to the work chamber.

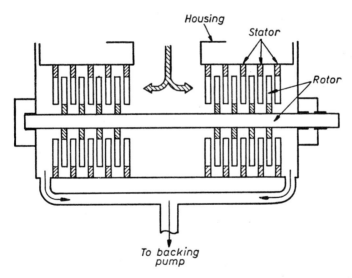

Figure 3.4   Turbo-molecular pump (schematic) (from *Vacuum* 1966, courtesy of Pergamon Press).

† Compare diffusion pumps, whose speed for lighter gases often exceeds that for heavier ones because the smaller thermal velocities, and hence diffusion, of heavy molecules causes a lower rate of entrainment (see § 3.3.1).

Each pair of discs forms a pressure stage, so that, with a backing pressure of 10^{-2} mbar, a partial pressure of air of order 10^{-10} mbar is achieved. With a speed of $140\,l\,s^{-1}$ the power consumption is about $\frac{1}{3}$ kW. The pressure ratio increases with the molecular weight of the gas pumped (hence the control of gear-oil backstreaming mentioned above). The fact that heavier gases are differentially pumped may be useful in some cases (e.g. in thermo-nuclear research), but in others (e.g. analysis of residual gases by mass spectrometry) it may be a source of difficulty.

The rotor may be mounted either horizontally (as in figure 3.4) or vertically, and may revolve at speeds as high as 60 000 rpm. Recently, a vertically mounted type with magnetically levitated bearings has been introduced, thus avoiding oil lubrication, and giving a pump free from organic vapours without the complication and loss of speed associated with LN$_2$-cooled† traps. However, turbo-molecular pumps are expensive in terms of capital cost per litre per second of pump speed. This tends to limit their use to applications where oil vapours cannot be tolerated. In Becker (1966, pp 625 ff) pumps are described with speeds for air ranging from about 140 to $5200\,l\,s^{-1}$ and pumps with speeds up to $9000\,l\,s^{-1}$ are now available.

A review of turbo-molecular pumps is given by Holland *et al* (1974).

3.3 Vapour-stream Pumps (Diffusion Pumps)

3.3.1 Basic principle

A liquid of high molecular weight boils under reduced pressure (provided by the 'backing pump') and by suitable geometry is formed into a vapour jet into which the gas molecules to be pumped diffuse, and are carried by the vapour jet towards the backing pump, whence they are expelled to the atmosphere. Pumps based on this principle are commonly called 'diffusion' pumps, since the molecules to be pumped diffuse into the jet of pump fluid vapour. If a stream of vapour molecules expands through a nozzle, it attains supersonic velocity if the pressure in the region into which it expands is sufficiently low. A number of nozzles are arranged in series so that each jet produces a reduction in pressure and provides the low pressure into which the succeeding jet expands.

The construction is shown schematically in figure 3.5, in which the pump fluid, boiling under reduced pressure from the backing pump,

† LN$_2$ is a common abbreviation for 'liquid nitrogen'.

passes up the cylindrical chimney, some of it emerging from the annular nozzle 3 and carrying with it gas which has diffused from stage 2 above it, thereby making the pressure at B less than that at A. The actions of nozzles 2 and 1 are similar, the reduction of pressure by a nozzle being typically several orders of magnitude.

The pump fluid must be removed and returned to the boiler after it has performed its function, hence the water-cooling of the pump casing. As each stage must handle the same number of molecules and the pressure decreases from A to D, the annular space between the nozzle and the pump casing increases in the same direction.

Diffusion pumps are often fitted immediately above the inlet port with a baffle which is intended to cause molecules of pump fluid migrating from the pump to the work chamber (i.e. going in the 'wrong' direction) to strike the baffle, be condensed upon it and return as liquid to the boiler. Thus the entry of pump fluid into the work chamber is minimised.

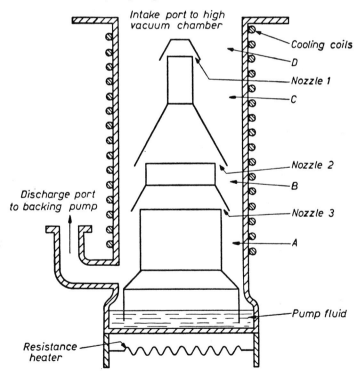

Figure 3.5 Vapour-stream diffusion pump (schematic).

Baffles, like cold traps (see § 2.6), are a compromise between opacity to the pump fluid going to the work chamber and opacity to the gas molecules coming from the work chamber, the latter being an undesired penalty. They are usually water-cooled, or refrigerated to about $-30\,°C$, and approximately halve the speed of the pump to which they are fitted. Baffles very deeply refrigerated with the intention of eliminating as completely as possible the passage of fluid from the pump to the work chamber are, in effect, cold traps.

3.3.2   Mercury diffusion pumps

The outstanding advantage of this (the earliest type of diffusion pump) is that the working fluid, being an element, cannot decompose in the boiler (in which the temperature may be about $150\,°C$). A major disadvantage is the high vapour pressure at the temperature of the cooling water ($\sim 10^{-3}$ mbar at $15\,°C$), so that unless a cold trap is used, the work chamber is flooded with mercury vapour. However, using a trap impedes the passage of gas between the work chamber and the top nozzle, and some loss of pump speed results. Mercury-filled pumps are, however, particularly suitable for systems in which organic contamination must be completely avoided and the use of sputter ion pumps (see § 3.4.1) is not desired. For details see Power (1966, pp 55 ff). Three-stage mercury pumps are available and can work against backing pressures as high as 25 mbar, so that backing by a 'filter' water-jet pump is possible.

3.3.3   Oil diffusion pumps

The working fluid is an organic compound having a vapour pressure at cooling water temperature in the range $10-10^{-4}$ mbar, depending on the species, and boiling under a backing pressure of a few tenths of a millibar in the temperature range $150-250\,°C$. For many applications, the presence of low residual oil pressures is acceptable and the use of a refrigerated vapour trap, with its attendant inconvenience and throttling of pumping speed, is unnecessary. If a cooled vapour trap is interposed between the pump and the work chamber, pressures of less than 10^{-11} mbar can be attained.

The disadvantage of organic pump fluids is that they tend, in service, to decompose into components whose room-temperature vapour pressure is much higher than that of the parent fluid.

The term *oil* to denote the working fluid is a loose one and covers a variety of organic fluids including paraffinic hydrocarbons, silicones, polyphenyl ethers and perfluoropolyethers. These have mass numbers of order 400, except the perfluoropolyethers with a mass number of about an order of magnitude greater than this. The significance of mass number is that both the collision area and the average momentum at a given temperature increase with the molecular mass. The relative merits of various vapour pump fluids are discussed by Laurenson (1980) who also gives useful information on rotary pump oils.

The choice between mercury and an organic fluid depends on the particular use contemplated. The pros and cons are discussed by Power (1966, pp 48 ff). According to Bentley (1980) the capital cost per unit pumping speed of oil diffusion pumps is less than that of any other type of high-vacuum pump.

A rough generalisation is that an oil diffusion pump is preferable to a mercury one unless there are some special considerations favouring mercury, such as the need to be completely free from organic vapours.

Oil booster pumps are similar in principle to oil diffusion pumps, from which they have been developed by increasing the backing pressures and heater input and thereby obtaining a greater through-put at the expense of ultimate vacuum. They are often employed to back large diffusion pumps, in order to 'boost' the discharge to a greater pressure before passing it on to a rotary backing pump. Their operating range is from about $1-10^{-3}$ mbar, filling the gap in which the rotary pump efficiency has begun to fall off while the diffusion pump has not yet attained its maximum.

3.3.4 *The speed of diffusion pumps*

In § 2.7, the speed S of a pump is given as the ratio of the net through-put Q to the pressure P at the entry port, i.e.

$$S = Q/P. \tag{3.1}$$

However, P must be unambiguously defined, there being two conventions with the difference between them becoming significant at low pressures. The pressure at the entry port of the pump is the sum of the partial pressure due to the pump fluid P_u (called its ultimate pressure) and the partial pressure P_g of the gas being pumped, i.e. the total pressure P_0 at the entry port is

$$P_0 = P_g + P_u. \tag{3.2}$$

The value of speed obtained by putting $P = P_g$ in (3.1) may be called the test-gas speed S, and there is no reason to suppose that in a well designed multi-stage diffusion pump $\dot{Q}$ should not remain proportional to P_g at very low pressures, so that the test-gas speed, denoted by S, remains constant. Alternatively, if P_0 is inserted in the denominator of (3.1), the speed S' measured in practice is obtained. Since the test-gas flow $\dot{Q}$ is independent of how P is defined, we have

$$\dot{Q} = S'P_0 = SP_g \tag{3.3}$$

which, inserting (3.2), becomes

$$S' = S\left(1 - \frac{P_u}{P_u + P_g}\right). \tag{3.4}$$

Hence, as the partial pressure (P_g) of the gas being pumped tends to zero, S' tends to zero, but S remains constant.

At the high-pressure end of the inlet pressure range, the speed (there being here no significant difference between S and S' since $P_g \gg P_u$) also decreases, probably because the critical backing pressure of one of the stages of the pump is exceeded. Figure 3.6, which illustrates the above phenomena, is due to Power (1966, pp 48 ff), whose treatment is followed here.

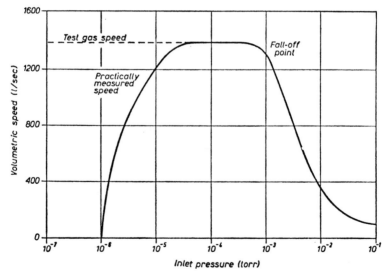

Figure 3.6   Test-gas speed and practically measured speed curves for a diffusion pump of area about 410 cm (from Power B D, *High Vacuum Pumping Equipment*, courtesy of Chapman and Hall Ltd).

The through-put of a pump, being its rate of removing gas, measured in *PV* units per unit time, is given at each inlet pressure by the product (pressure × speed). This product is a maximum at the high-pressure end of the speed/pressure curve, usually in the 10^{-2} mbar region.

In § 2.7, it is stated that the maximum speed of a pump for air at 25 °C cannot exceed 11.9 l s^{-1} per cm^2 of entry port. In most diffusion pumps only about $\frac{1}{3}$ or $\frac{1}{4}$ of the molecules which cross the mouth entry plane are entrained in the vapour stream never to return, the proportion so entrained being called the 'Ho coefficient', which (for example) in figure 3.6 has the value (1350/410 × 1/11.9), i.e. about 0.28 at 10^{-4} Torr.

3.3.5 Backstreaming of diffusion pumps

Backstreaming is a measure of the tendency of a pump to contaminate the work chamber it is pumping with its oil. Some of the molecules issuing from the top jet find their way into the 'wrong' direction, i.e. upwards towards the work chamber instead of downwards towards the jets beneath.

A common measure of backstreaming is the rate in mg cm^{-2} min^{-1} at which this unwanted oil enters the work chamber. The area through which the oil flux passes is taken as the mouth area of the pump, though the flux is not uniform over the mouth, but tends to be annular in distribution. Backstreaming is reviewed by Harris (1977), who discusses the main sources and the methods used to control them. He quotes values ranging from 0.5–8.3 × 10^{-5} mg cm^{-2} min^{-1}.

All pump fluids except mercury are to some extent liable to decompose when heated in the boiler of the pump, and may contribute to the total backstreaming. The stability of various pump fluids has been reviewed by Laurenson (1980).

3.3.6 Steam ejector pumps

In these, a high-speed jet of steam entrains the gas to be pumped, and ejects it at a higher pressure. In contrast to oil or mercury diffusion pumps, the working fluid (steam) is not condensed in the pump itself, but leaves it, moving in the same direction as the ejected (i.e. compressed) gas. Up to seven stages have been used in series and ultimate pressures of 10^{-1} mbar and lower can be reached, the high-pressure stage ejecting straight to the atmosphere. Steam ejector pumps are particularly useful in large-scale industrial plants where steam and cooling water are already available, the pumping loads are 'dirty', and oil vapour-stream pumps are therefore unsuitable. For details, including performance and economic considerations, see Beck (1964).

3.4   Chemical Pumps

Chemical pumps are so called because the gas to be pumped forms a chemical (and generally irreversible) bond with an absorbent in the pump. This distinguishes them from sorption pumps, which rely on van der Waals adsorption between gas and adsorbent, and from cryopumps, which depend on bonds between like molecules deposited on the cold walls of the pump. All three types of pump have the common property of retaining the pumped gas molecules, as opposed to expelling them.

3.4.1   Sputter ion pumps

In sputter ion pumps there is a cold-cathode discharge (see § AII.17) in the gas to be pumped, between an anode and cathodes of a highly reactive metal, usually titanium. The impact of the positive ions on the cathodes sputters (see § AII.18) the titanium on to surfaces of the pump, where the gas to be pumped reacts with it and is immobilised. In addition, some of the positive ions are driven into the cathode and are 'buried' there. The latter action is especially important for the noble gases, which do not react chemically with titanium.

Figure 3.7 shows schematically the geometry of one simple design, the cathodes being titanium and the anode stainless steel. Electrons emitted from the cathodes oscillate in a potential well between them, produced by the positive potential of the anode, and prior to an ionising collision are constrained by a magnetic field (of the order of 1–2 kG) to move in spiral paths, thus increasing the ionisation probability. The more massive positive ions, whose motion is not seriously affected by the magnetic field, strike the cathodes and sputter titanium onto the walls of the anode cells. Ions which are buried in the cathode are liable to disinterment by subsequent ionic bombardment, and in sophisticated designs this possibility is minimised by non-planar cathode geometries (see Lewin 1965). The disinterment phenomenon is particularly troublesome with argon, which constitutes about 1% of atmospheric air, and leads (except in pumps specially designed to avoid it) to argon instability, in which argon is periodically released from the cathodes, causing step-like rises in pressure of as much as two decades. For a detailed account of sputter ion pumps, see Robinson (1968).

Sputter ion pumps are particularly indicated when a system free from oil vapour is desired, and hence they are often pumped down to the starting pressure (10^{-2} mbar or less) by a sorption pump (see § 3.5) or a trapped rotary pump. To get easy and reliable starting, certain precautions are necessary, the most important being to pre-pump to as low a pressure as practicable and to do so by means of an adequately sized connection to the starting pump. The pumping speed is a function of pressure and of previous history; speeds between $10\,\mathrm{l\,s^{-1}}$ and $10\,000\,\mathrm{l\,s^{-1}}$ are available.

The advantages include low power consumption, independence of orientation with respect to the vertical, quietness (no mechanical backing pump needed), freedom from attention (no disastrous consequences of power failure), utility as its own leak detector (see § 5.10.4) and the ability to produce, if required, pressures of 10^{-10} mbar or less in an oil-free system.

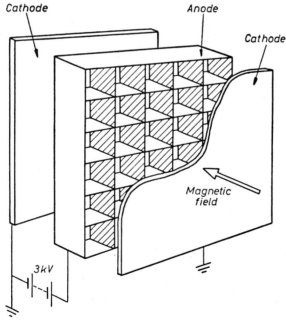

Figure 3.7   Sputter ion pump (diagrammatic)
(courtesy of Mullard Research Laboratories Ltd).

Since the gas pumped is not ejected, the useful life is inversely proportional to the pressure, but at low pressure this is not a serious limitation, e.g. at 10^{-10} mbar a $100 \, l \, s^{-1}$ pump will take about three years to become saturated.

3.4.2   Titanium sublimation pumps

As a freshly deposited film of titanium chemically combines with nearly all gases except the noble ones and methane, any means of producing such a film on the inner surface of a vessel can form the basis of a pump. One such means involves a water-cooled anode supporting a pendant drop of titanium, which is heated to evaporation temperature by electron bombardment from a thermionic cathode, and continuously fed by a titanium wire. The evaporated titanium is condensed on water-cooled surfaces.

A simple method, with limits, however, to the pumping life, involves sublimation of titanium from an electrically heated filament of titanium/molybdenum alloy, and its condensation on water- or LN_2-cooled walls, the sticking probability of the gas molecules (i.e. the chance of sticking to the Ti film on the first impact) being considerably greater on the cooler surface.

Sublimation pumps can be started at pressures of 10^{-3} mbar or less (obtained by other means) and they are particularly useful when operated in short bursts to reduce pressure from (say) 10^{-8} to 10^{-10} mbar. Titanium sublimation pumps can have speeds reaching many thousands of litres per second, since, with suitable ('crab pot') geometry, the speed factor of the entry port (see § 3.3.4) can approach unity. The pros and cons of this type of pump have been summarised by Power (1966, pp 310 ff). Small, expendable glass sublimation pumps are available which consist of an ordinary Bayard–Alpert pressure gauge (see § 4.5.1), into which an extra titanium getter or zirconium-coated tungsten filament has been introduced. The system is evacuated to (say) 10^{-7} mbar by conventional means, the getter filament then being heated and the pressure reduced to about 10^{-10} mbar. Prior to its use as a getter pump, the gauge can be used for pressure measurement in the ordinary way.

3.4.3   Other chemical pumps

A good getter material for a pump should be chemically active when freshly distilled, have a low vaporisation rate at the maximum temperature at which vacuum vessels in which it is used are degassed, but evaporate readily at higher temperatures, and should not react unduly when exposed to the atmosphere.

Barium, alloyed with aluminium to produce a protective oxide film when exposed to the atmosphere, is often used as a getter in electronic tubes. The mechanism of its uptake has been studied by Bloomer and Cox (1965). Numerous other metals have been used for special purposes, e.g. the pumping of H_2 and D_2 by evaporated Mo has been described by Hunt *et al* (1961), and the use of Al/Zr getters for handling tritium by Lange (1965). A summary is given by Power (1966, pp 287 ff).

3.5 Sorption Pumps

Sorption pumps rely on physical adsorption between unlike molecular species, namely the adsorbent surface and the gas molecules to be pumped. The mechanism essentially consists of the formation of a monolayer† (or less), since in n-layer adsorption (where n exceeds 2 or 3) the vapour pressure of the adsorbed layer approximates to that of the adsorbent in bulk, and the action is more properly termed cryopumping—to be discussed in § 3.6. The efficiency of sorption pumping is determined by a competition between the adsorption energy q, tending to hold a gas molecule on the adsorbent surface, and the thermal energy kT tending to dislodge it, the determining factor in the escape probability being $\exp(-q/kT)$ where k is Boltzmann's constant, and T the absolute temperature of the adsorbing surface (see § AII.19).

Hence, a good adsorbent should have a large accessible area per unit volume (implying porosity) and a value of q, such that $\exp(-q/kT)$ increases by a large factor when T is raised from a conveniently low working temperature (e.g. LN_2 at 77 K) at which the surface coverage is large, to the highest activation (i.e. degassing) temperature which does not permanently impair its low-temperature adsorbent properties. Two commonly used adsorbents are charcoal and artificial zeolites. The porosity of the former derives from the cellular structure of the original wood, and that of the latter from the molecular sized cavities left by dehydration, a process which does not involve the collapse of their original cavity structure.

The porosity of charcoals is such that, if the pores were cylindrical,

---

† See § AII.5.

they would be of diameter about 50 Å and on this basis the internal area of 1 cm^3 of a good activated charcoal would be (see Dushman 1962) about 10^3 m^2.

The zeolites are three-dimensional Si—O—Al anionic networks, containing interstitial and exchangeable cations and neutral water molecules, the latter being removable by heating without structure collapse, leaving cavities of atomic dimensions. The size of the cavities varies from about 3–10 Å, depending on zeolite type. The zeolites adsorb selectively according to the relation between their cavity size and the size of the molecule to be sorbed, hence the name *molecular sieves*. The equivalent internal surface of the synthetic zeolites is about 600–800 m^2 g^{-1}.

Zeolites are activated (i.e. freed from water and other adsorbates) by heating to several hundred degrees centigrade *in vacuo* or by flushing, when heated, with dry nitrogen. When activated under optimum conditions, one gram of a typical zeolite (Molecular Sieve 5A, manufactured by Union Carbide) will, when in equilibrium at 77 K with 1.3×10^{-3} mbar in the gas phase, carry a burden of about 100 cm^3 (measured at NTP) of nitrogen (see Stern and Dipaolo 1967), which is several orders greater than the burden under the same pressure at room temperature. However, the pre-adsorption of 5% of water by weight will reduce the burden it will carry at 10^{-3} mbar and 77 K to about 20 cm^3 measured at NTP. At 77 K the zeolites adsorb useful amounts of most common gases. H$_2$ is little adsorbed, and Ne and He hardly at all. The presence of the two noble gases is a limiting factor in the vacua which may be obtained starting from air at atmospheric pressure. The adsorption of zeolites increases significantly with the number of previous sorption and reactivation cycles they have undergone, but is very adversely affected by pre-adsorption of water, which is one of the most difficult of the common gases to remove. If the adsorption of water is avoided, reactivation necessary for effective pumping of dry air only involves letting the pump warm up to room temperature after use at 77 K.

The zeolites are usually in the form of pellets, of millimetre dimensions, held together with 20% of clay. Since the thermal conductivity of zeolites is small, and since as soon as they produce vacua such that the mean free path is limited by inter-pellet distance the effective conductivity decreases, the design should be such that the maximum distance of any part of the zeolite from the refrigerant-cooled surface does not exceed (say) 1 cm.

The metal gauze cylinder shown in figure 3.8 facilitates access of the gas to be pumped to the zeolite-filled adsorbing annular space. The safety valve vents the gas desorbed to the atmosphere when the pump returns to room temperature.

Precise quotation of pumping speeds is not possible, but for easily adsorbed gases like O_2, N_2 and Ar, the speed of a well designed pump is often of the order of the conductance of its neck, implying that the chance that a molecule which enters the pump will escape sorption is small.

Though most commercially available pumps are designed for refrigeration by LN_2, lower temperatures are required for sorption pumping of H_2, Ne and He, and pumps have been designed and tested (see Stern *et al* 1965, 1966) using molecular sieves cooled with liquid H_2 (BP 20.3 K) or liquid He (BP 4.2 K). It is believed that such pumps may be useful in evacuating large space-simulation chambers to very low pressures.

A common use of LN_2-cooled sorption pumps is to provide the starting pressure (10^{-2} mbar or less) for sputter ion pumps in systems where complete freedom from the possibility of organic contamination is

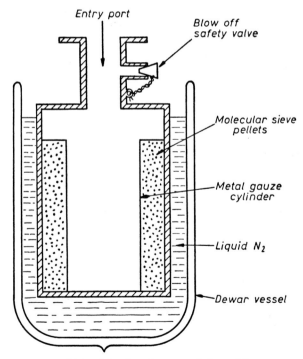

Figure 3.8 Sorption pump (schematic).

desired (see § 3.41). For example, Lange (1965): a work chamber of 20 l volume, previously flushed out with dry N_2 and evacuated by a LN_2-trapped water-jet filter pump to 25 mbar can be taken down by 85 g of LN_2-cooled zeolite 5A to 1.3×10^{-2} mbar in 14 minutes and to 6.7×10^{-3} mbar in 23 minutes.

3.6 Cryopumps

In cryopumps the gas being pumped is condensed onto the solid phase of the gas already pumped, so that a thickness of many molecular layers may be built up. For high efficiency, it is therefore necessary that the sticking coefficient of impinging gas molecules on the gas already condensed should be high and that the solid so built up should have a saturated vapour pressure smaller than the ultimate pressure the pump is desired to achieve.

The simplest cryopump is the LN_2-cooled vapour trap which is very efficient in pumping water, whose vapour pressure at 77 K is completely insignificant. For this reason the addition of a LN_2-cooled trap to an oil diffusion pump may be justified in situations where the advantage of its excellent speed for H_2O outweighs the disadvantage of decreased speed for permanent gases, due to decreased conductance to the diffusion pump.

However, at 77 K many common gases are not condensed, so that cryopumps are often cooled by LH_2 (BP 20.3 K) or LHe (BP 4.2 K). To minimise consumption of LH_2 or LHe and to prevent an excessive temperature rise of the surface of the condensed layer of gas above that of the underlying cryopanel, radiation shields, often cooled with LN_2, are arranged so that the cryosurfaces do not 'see', and therefore receive radiation from, the surrounding world at about 300 K.

The construction is shown schematically in figure 3.9. The condensing cryopanel, cooled by LHe, is surrounded by a radiation shield cooled by LN_2, the gases to be pumped reaching the cryopanel via cooled baffles which limit the entry of unwanted radiation without unduly decreasing the pumping speed.

Cryopumping is usually preceded by evacuation via mechanical or diffusion pumps to pressures of the order of 10^{-2} mbar both to economise in the use of refrigerant (since condensation of gas liberates latent heat) and to increase the useful life of the pump before the layer of condensed gas becomes so thick, and its thermal resistance therefore so large, that the condensing surface becomes significantly warmer than the underlying cooled cryopanel.

Cryopumps have high speeds in relation to their overall dimensions and, unlike diffusion pumps, need not be mounted vertically. In recent years, a self-contained form of cryopump has been available. In this, LN_2 and LH_2 are replaced by a small closed-cycle gaseous He refrigerator. The cryopanels are held at about 15 K, a temperature adequate to pump to a very low pressure all atmospheric gases except H_2, He and Ne. These are removed by sorption (see § 3.5) on charcoal or zeolites held at 15 K by an adhesive on the inside of the cryopanel.

Cryopumps are used in a number of special situations including the evacuation of large low-density wind tunnels, large space-simulation chambers and beam line pumping. In small laboratory apparatus a 'finger' cooled by liquid helium is useful in quickly obtaining a good vacuum, e.g. in experiments on clean surfaces.

Recent developments in cryopumping have been reviewed by Hands (1982). Bentley (1980) describes modern cryopumps and discusses their strengths and limitations, including economic considerations.

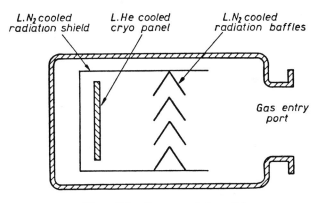

Figure 3.9 Cryopump (schematic).

References

Beck A H 1964 ed *Handbook of Vacuum Physics* vol 1 (Oxford: Pergamon) pp 77 ff
Becker W 1966 *Vacuum* **16** 625

Bentley P D 1980 *Vacuum* **30** 145–58
Bloomer R N and Cox B M 1965 *Brit. J. Appl. Phys.* **16** 1331
Dushman S 1962 *Scientific Foundations of Vacuum Techniques* 2nd edn, ed J M Lafferty (New York: Wiley) p 437
Hands B A 1982 *Vacuum* to be published
Harris N S 1977 *Vacuum* **27** 519–30
Holland L, Steckelmacher W and Yarwood J 1974 *Vacuum Manual* (London: E and F N Spon) pp 340–2
Hunt A L, Damm C C and Popp E C 1961 *J. Appl. Phys.* **32** 1937–41
Lange W J 1965 *J. Vac. Sci Technol* **14** 582–7
Laurenson L 1980 *Vacuum* **30** 275–81
Lewin G 1965 *Fundamentals of Vacuum Science and Technology* (New York: McGraw-Hill) p 156
Power B D 1966 *High Vacuum Pumping Equipment* (London: Chapman and Hall)
Robinson N W 1968 *The Physical Principles of Ultra-High Vacuum* (London: Chapman and Hall) pp 50 ff
Stern S A and Dipaolo F S 1967 *J. Vac. Sci. Technol* **4** 347–55
Stern S A, Mulhaupt J T, Hemstreet R A and Dipaolo F S 1965 *J. Vac. Sci. Technol* **2** 165–77
——1966 *J. Vac. Sci. Technol* **3** 99

4 Measurement of Pressure

4.1 Introduction

The state of affairs within a vacuum chamber is determined by the molecular species present, the number density n (number per unit volume) of each, and the temperature, which in equilibrium will be the same for each species.

The total pressure is the sum of the partial pressures of the constituents, and since

$$P = nkT \qquad (4.1)$$

for each species (see § AI.5) the total pressure P is determined by the temperature and the number density of the constituents. The pressure (in the sense of force per unit area) is seldom important in vacuum technology, the interesting quality usually being the number density n—a quantity measured by nearly all common gauges, except the McLeod gauge, which despite grave limitations is the standard calibrating instrument for many gauges.

This chapter will deal with the McLeod and its limitations, with some other common gauges and their calibration, with thermal transpiration, very briefly with mass spectrometers considered as instruments for determining relative number densities of various constituents, and with sorption/desorption methods.

In technology, it is often sufficient that the reading of the pressure gauge does not differ significantly from a value known from previous experience to be acceptable, and so provided they are reproducible the inaccuracy of gauge indications is not necessarily serious. In some scientific work however, e.g. investigation of chemical gas kinetics at low pressures, the exact numerical value of the number density may be desired. Unless otherwise stated, vacuum gauges are usually calibrated in N_2 pressure, and the pressure of other gases obtained with them quoted as 'nitrogen equivalent'.

4.2   Survey of Types of Gauge Measuring Total Pressure

4.2.1   Barometric types

The liquid may be mercury or oil. Without instrumental aid, the minimum difference of level which can be detected is of the order of 10^{-1} mm, so mercury types have a lower useful limit of about 1 mbar if 10% accuracy is required, the sensitivity of oil types being about 15 times greater. In some situations, the considerable improvement of sensitivity afforded by instrumentation (optical or electronic) may be justified.

4.2.2   Mechanical types

The most usual type is a capsule gauge in which the pressure causes mechanical movement of a septum forming one wall of an evacuated enclosure. The principle is similar to that of the aneroid barometer. Models are available in various ranges, e.g. 0–25, 0–50, 0–125 and 0–1000 mbar. The sensitivity of gauges of this type may be considerably increased by suitable instrumentation.

4.2.3   Gauges of the McLeod type

The basic principle is to take a sample of the gas, decrease its volume by a known factor and measure the resultant pressure. The manometric liquid is usually mercury; details are given in § 4.3. It assumes Boyle's Law, i.e. *PV* is constant at constant *T*.

4.2.4   Thermal conductivity gauges

These depend on the decrease of thermal conductivity at low pressures (see § AI.10) resulting in an increase of the thermal insulation of a heated body as the pressure is reduced. Details are discussed in § 4.4.

4.2.5   Hot-cathode ionisation gauges

In these the number density is deduced from the amount of ionisation produced by the passage of a known current of electrons of fixed initial energy through a fixed geometrical structure. Details are given in § 4.5.

4.2.6 Cold-cathode ionisation gauges

In these, the pressure is deduced from the current in a cold-cathode discharge tube under controlled conditions of applied voltage. Details of the Penning gauge, which utilises this principle, are given in § 4.6.1. Useful qualitative indications are given by an HF discharge as mentioned in § 4.6.2.

4.2.7 Gauges depending on other physical properties

These include gauges utilising the decrease of velocity at low pressures (analogous to the decrease of thermal conductivity utilised in the thermal conductivity gauges), and the Knudsen gauge, which depends on the recoil force on a warm surface when a molecule rebounds from it (see §§ 4.7 and 4.8).

The most common gauges are mechanical gauges of the diaphragm type, thermal conductivity gauges, hot-cathode gauges and cold-cathode gauges, while the McLeod is much used for calibration.

4.3 The McLeod Gauge

The general principle of operation, stated in § 4.2.3, namely compression of a sample gas by a known factor and measurement of the resultant pressure, is valid only if no significant constituent of the gaseous mixture is caused by the compression to condense to liquid or be adsorbed by the walls of the gauge.

Figures 4.1(*a*) and (*b*) show alternative methods of operation, the geometries of the glass tubing being shown, for simplicity, as identical. An auxiliary vacuum is applied to the reservoir R, so that the mercury level falls below P and the bulb B is filled with gas at the pressure p_0 to be measured. Atmospheric air is then slowly admitted to R so that the mercury rises and, on passing P, traps a sample of gas in the volume (*V*) between P and the closed end of the capillary (cross section area *S*) which is sealed to the bulb B. The subsequent procedure differs for the two modes.

In the mode of figure 4.1(*a*) the mercury is allowed to rise until it reaches a fixed mark at level *a–a*. The gas compression ratio is then V/ls and its pressure is therefore $p_0 (V/ls)$, where *l* is the distance in millimetres between *a–a* and the level *c–c* of the top of the capillary

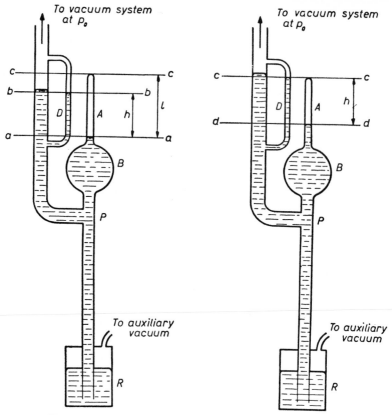

Figure 4.1 McLeod gauge; (*a*) linear scale, (*b*) square-law scale.

tube. The level *b–b* to which the mercury rises in the limb D (which, to minimise uncertainties due to surface tension, is of the same bore as A) is noted. If *h* is the vertical distance in millimetres between *b–b* and *a–a*, the total pressure at *a–a* is $p_0 + h$ mm Hg, i.e. $p_0 + h$ Torr. Hence

$$p_0 + h = p_0 V/ls$$

or

$$p_0 = \frac{h}{V/ls - 1}$$

i.e.

$$p_0 \approx \frac{lsh}{V} \text{ mm Hg (Torr)†} \tag{4.2}$$

† The scale can be calibrated in millibars, since 1 millibar equals $\frac{3}{4}$ of a torr.

since in practice $V \gg ls$. Thus a scale fixed behind D, from which the height h is measured, is linear in p_0.

When operated as shown in figure 4.1(b), air is admitted to R until the mercury in D has reached the level c–c of the tip of the capillary C and the distance h mm between c–c and d–d is read. The compression ratio is V/sh and the pressure p_0 is given by

$$p_0 + h = p_0 V/sh$$

therefore

$$p_0 = \frac{h}{V/sh - 1} \tag{4.3}$$

or

$$p_0 \approx \frac{sh^2}{V} \tag{4.4}$$

since

$$V/sh \gg 1.$$

Thus the pressure, which is read on a scale fixed behind A, is proportional to the square of h.

The advantage of the mode shown in figure 4.1(a) is linearity of scale, so interpolation by eye between scale markings is easy. On the other hand, the compression ratio is fixed, while in the mode of figure 4.1(b) it is variable, increasing as the pressure becomes lower.

The practicable compression ratio (i.e. pressure amplification) is determined by (i) the maximum weight of mercury which can be conveniently supported in a glass bulb (the volume V) and (ii) the minimum allowable value of S, which increases with the surface tension depression and the 'stiction' of the mercury meniscus in D and A, and varies with the cleanness of the mercury and the surface state of the glass. These factors may not be the same in D and A, even though they are nominally identical, and thus become serious at small diameters. The advantage of a compression ratio which increases as the pressure decreases, and the consequent improvement of the sensitivity at low pressure and larger useful working scale, has made the 'square-law' McLeod gauge the more popular instrument.

To get some 'feel' for magnitudes, consider a square-law gauge with $V = 10^2$ cm^3 (corresponding to nearly $1\frac{1}{2}$ kg of mercury) and $S = 10^{-2}$ cm^2 (i.e. a capillary diameter of order 1 mm) used to measure a pressure of 1.3×10^{-2} mbar. Then from equation (4.4), $h = 10$ mm, and if the error in reading h is 0.5 mm of scale length, the error in p_0 is about 10%.

The above example is not intended to suggest that greater accuracies cannot be achieved by using larger V and smaller S. However, for good accuracy, special care must be taken, especially in the surface treatment of the inner surface of the capillary and the uniformity of its bore. It is probably reasonable to assume that unless special precautions are taken, the gauge is not reliable as an absolute standard below about 10^{-4} mbar. Leck (1964, pp 3ff) should be consulted for details.

4.3.1   *The mercury vapour-stream effect*

Unless isolated by an intervening cold trap, a McLeod gauge will contaminate the vacuum system to which it is connected with mercury vapour, to a degree dependent on the temperature and therefore the vapour pressure of the mercury. However, the cold trap, in protecting the system from mercury vapour, introduces another error. The vapour coming from the McLeod and condensing in the trap produces a pumping effect, causing the gas pressure indicated by the gauge to be less than that in the system on the side of the trap remote from the gauge.

The magnitude of the effect increases with the diameter of the pipe connecting the gauge and the cold trap, and with the temperature and therefore the vapour pressure of the mercury. Since the inequality of gas pressure set up by the stream of mercury vapour moving from the gauge to the trap is opposed by the diffusion of gas in the opposite direction, the effect is less the greater the diffusion coefficient of the gas through the mercury vapour. The resultant error for the rare gases and for N_2 and CO_2 has been measured by Elliott *et al* (1967), whose paper includes useful references to the literature. They state that the effect is proportional to the tube diameter and, with a tube of effective diameter 1 cm and a temperature of 23 °C, is about 2% for He and 27% for Xe over the pressure range 4×10^{-6}–1×10^{-4} mbar. The error can be reduced by refrigerating the McLeod (Rothe 1964), since Hg remains liquid down to -38.8 °C, and by reducing the bore of the tube between the gauge and the cold trap. For necessary precautions in the latter case see Tunnicliffe and Rees (1967).

4.4   Thermal Conductivity Gauges

In general the heat loss from a hot body placed in a vessel with cooler walls depends on geometry and the thermal conductivity of the gas.

However, as the pressure is reduced, a stage is reached at which the thermal conductivity decreases with pressure and indeed the term itself loses its usual meaning. Consider a hot wire centrally placed in a vessel with its walls at room temperature. At a sufficiently high pressure, there will be a temperature gradient from the wire to the wall and the situation may be expressed in terms of geometry and a constant thermal conductivity (see e.g. Roberts and Miller 1955). When the pressure is so low that the mean free path is large compared with the wire diameter, a molecule which has hit the wire and been heated will subsequently make many collisions with the other molecules and with the walls of the vessel before hitting the wire again. Hence the temperature of the molecules hitting the wire will be approximately that of the walls. In these circumstances, the rate of heat loss decreases linearly with the bombardment rate, i.e. pressure.

This mechanism governs the behaviour of the three common types of thermal conductivity gauge—Pirani, thermocouple and thermistor—now to be described. In all three types, the temperature of the detecting resistance element is low enough to allow it to operate in oxidising gases without damage.

4.4.1 The Pirani gauge

The basic principle of this gauge is that the heat loss from a current-carrying wire, diameter d, mounted in an envelope connected to the work chamber decreases with pressure if $\lambda \gg d$, where λ (see §§ AI.9 and AI.10) is the mean free path.

The principle has been applied in two ways: (*a*) keeping the temperature of the wire constant as the pressure changes, the necessary voltage change being a measure of change of pressure, or (*b*) allowing the temperature of the wire to change as the pressure changes, and observing the resistance change of the wire.

Method (*b*) is more common, the wire being one arm of a Wheatstone bridge and the pressure change being indicated by change in the out-of-balance current in the bridge microammeter.

Figure 4.2 shows the essentials of a constant-voltage bridge, the arm R_1 being a tungsten wire in a glass envelope connected to the work chamber whose pressure P it is desired to read on M. To zero the instrument, the pressure in the envelope of R_1 is reduced to a value such that the heat loss *via the gas* is negligible compared with that via radiation and thermal conduction from the tungsten to its leads, the potentiometer R_5 being adjusted to make the reading of the meter M zero. If gas is now introduced into the envelope of R_1, the tungsten filament is cooled, the bridge is unbalanced and a meter reading is produced which is a function of P.

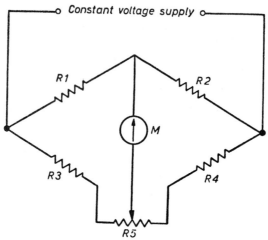

Figure 4.2   Constant-voltage Pirani bridge.

To compensate for changes of ambient temperature the temperature coefficient of R_2 is chosen to be approximately equal to that of R_1, though as Leck (1964) points out, the compensation is only effective over a limited pressure range. At a given gas temperature and pressure, the rate at which the molecules bombard the filament is proportional to (molecular weight)$^{-1/2}$ (see § AI.6) and the amount of heat carried away per impinging molecule depends on the difference between the temperatures of the wire and the gas, the specific heat of the gas, and its thermal accommodation coefficient α (see § AII.6) on the tungsten filament†.

Thus the calibration of a Pirani, usually done for N_2 or air, depends on the gas. For example, a gauge calibrated for N_2 will read about 60% high for H_2 at the same pressure, because H_2 is the more efficient coolant.

A useful arrangement which doubles the sensitivity is to make all four arms of tungsten, R_1 and R_4 being in a single bulb which is at the pressure P to be measured, and R_2 and R_3 being mounted in a similar

---

† Fortunately, in the relatively high-pressure region in which Piranis are used, adsorbed gases make α approximately constant and not far from unity.

bulb which is evacuated and permanently sealed off. The two bulbs are put close together, preferably in a thermally insulating enclosure, to compensate for changes in ambient temperature. Numerical magnitudes of a typical arrangement are as follows. R_1, R_2, R_3 and R_4 are tungsten filaments of about 0.05 mm diameter, with a room temperature resistance of order 20 Ω, which attain a temperature of about 300°C when a stabilised 2 V supply is applied to the bridge. A 0–200 microammeter, with a resistance of the order of 200 Ω, gives a full-scale deflection corresponding to a N_2 pressure of 4×10^{-3} mbar. The minimum useful reading is about 10^{-4} mbar. By shunting the meter, three or more ranges going up to (say) 10^{-1} mbar can be obtained with this arrangement, but linearity is sacrificed at the higher pressures, the instrument becoming increasingly insensitive.

However, by special design (see Cleaver 1967) instruments with a useful range of up to 1 mbar are possible, but above this pressure other types of gauge, e.g. the diaphragm gauge, are usually preferable. It is possible to construct considerably more sensitive Piranis, but these are subject to excessive zero drift due to causes such as changes in ambient temperature and in radiation and gas thermal accommodation surface properties of the filaments.

By deliberate exploitation of gas convection as opposed to conduction, thermal conductivity gauges may be used up to atmospheric pressure; see e.g. Steckelmacher (1965), who gives a useful account of pressure measurement from an industrial viewpoint.

The speed of response of a constant-voltage Pirani gauge is determined by the thermal capacity of the W filament, and the degree of thermal linkage between the filament and the gas; the time constant is usually of the order of some seconds. However, by using a filament of diameter 10^{-2} mm, maintained at a constant resistance by feedback circuits, a time constant can be attained which is 0.14 s at 10^{-6} mbar and 0.02 s at 10^{-1} mbar (see Allsopp *et al* 1969).

4.4.2 The thermocouple gauge

The principle, identical with that of the Pirani, is the variation of heat loss from an electrically heated wire in the pressure region where it decreases with decrease of gas pressure.

One simple form consists of a platinum ribbon heated by a constant current to between 100°C and 200°C, the temperature (which is a function of pressure) being measured by a thermocouple attached to it, and the output read on a sensitive microammeter.

The scale is non-linear and the lower limit of use is about 10^{-3} mbar. Most commercial instruments cover the range 10^{-3}–1 mbar, but other ranges can be covered by appropriate designs.

The advantages of this gauge are ruggedness, simplicity and (as in the Pirani gauge) the use of a filament running at a low temperature, which is therefore not injured by exposure to atmospheric air. It is also useful in detecting sudden and accidental pressure rises, and activating the necessary 'shutdown' devices. Like the Pirani, its calibration is a function of gas species.

4.4.3   The thermistor gauge

This depends upon the heat loss from a small bead of semiconducting material fastened to two leads, and enclosed in a small bulb connected by tubulation to the vacuum work chamber. The material has a large negative temperature coefficient of resistance, and is self-heating when a current is passed through it. An increase in gas pressure causes an increase in cooling, a decrease of temperature and an increase of resistance and of voltage drop (see Dushman 1962, pp 297 ff).

The calibration, as in the other thermal conductivity gauges, depends on the gas, the reading being higher, for example, for H_2 than for N_2 at the same pressure. The extension of the upper limit of measurement of a thermistor gauge, by surrounding the thermistor bead with an electrically non-conducting powder, is described by Green and Lee (1966).

4.5   Hot-cathode Ionisation Gauges

These gauges depend (see §§ AII.11 and AII.12) on the thermionic emission of electrons from a hot filament, which are accelerated by an electric potential to a velocity which maximises their probability of ionising the gas molecules which they hit. The positive ions so produced are collected by an electrode whose potential is such that it cannot collect electrons which, if collected, would diminish the current attributed to the positive ions.

The rate of ion production is proportional to the number density of the gas molecules and to the number of ionising electrons emitted per second from the hot filament. Hence, for a given ionising electron current i_g, the positive ion current i_c produced is proportional to the number density of gas molecules n, which is related (see equation AI.5) to the pressure P by the relation

$$P = nkT. \tag{4.5}$$

The characteristics of an ionisation gauge are therefore expressed by the equation

$$i_c = Gi_gP \tag{4.6}$$

where G is called the gauge sensitivity, and is specified as amps per amp per torr, i.e. the sensitivity is given in $mbar^{-1}$, and is typically of order 10. Gauge sensitivities are usually quoted for N_2, room temperature being implied; the specification of temperature is relevant, since the gauge really measures n, and not P.

The accelerating voltage applied to the electrons from the filament is usually between 100 and 200 V for the reason given in § AII.11.

The underlying assumptions and the range of validity of equation (4.6) are discussed by Dushman (1962, pp 301 ff).

The use of a hot-cathode ionisation gauge to measure gas pressure involves the possibility of several undesired side effects. One effect (which causes the gauge to read high) is the emission of adsorbed gas from its electrodes and walls under the influence of heat and/or electrons from the filament. The remedy is to degas all parts of the gauge at temperatures higher than will be attained in service (see Leck 1964, p 91). An effect which may cause too low a gauge reading is the reaction of the hot filament with the gas. For instance O_2 reacts with hot W producing an oxide which is deposited on the gauge walls (see Leck 1964, p 103). Also, positive ions or molecules excited by electron impact may react with the walls of the gauge (see e.g. Lewin 1965) which becomes in effect a pump.

All these effects are minimised by using a filament of minimum area, run at as low a temperature as possible, using the smallest practicable value of i_g, and connecting the gauge to the work chamber by tubulation of as large a bore as possible. The limiting case of the latter is to mount the gauge element in the work chamber itself, the so called 'nude gauge' technique.

4.5.1 The Bayard–Alpert gauge

Figure 4.3 shows the essential features of a BA gauge, the thermally emitted electrons from the hot filament (usually W) being accelerated to the grid. They make a few oscillations through its wires and are finally captured by it, constituting the i_g of equation (4.6). Before their final capture, a small fraction of these electrons ionise gas molecules, and most of the ions produced inside the grid are collected by the thin wire in its centre and constitute the current i_c. This current, amplified externally as necessary, is, for a given value of i_g, a direct measure of the number density n, or gas pressure P.

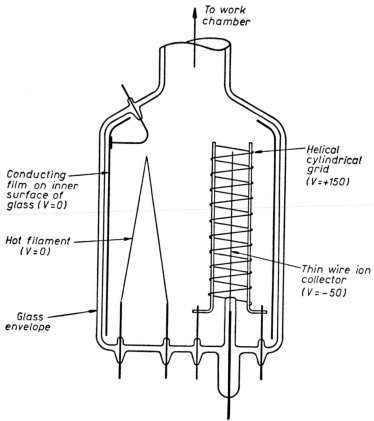

Figure 4.3 Bayard–Alpert ionisation gauge (schematic) (figures in brackets are volts with respect to filament).

Stable sensitivity demands stable potentials within the gauge, so it is advantageous to control the potential of the inner surface of the glass envelope by means of a thin conducting film. Cobic *et al* (1961) have investigated the occurrence of a bistable mode of operation of a BA gauge, due to envelope potential.

The electrons from the filament, on striking the grid, generate soft x-rays (see § AII.15) which in turn cause the emission of photoelectrons from the collector, thus increasing the *apparent* ionisation current i_c, and to that extent falsifying the readings of the gauge.

In the BA gauge, this soft x-ray effect does not become serious (i.e. comparable with the ion current) until the pressure falls to about 10^{-10} mbar. The collector is a *thin* wire (and hence has a small target area for photons) in order to achieve this. Earlier types of ionisation gauge, in which the electrode arrangement is different and the collector much larger, have a lower limit of pressure of about 10^{-7} mbar, due to the soft x-ray effect. The sensitivity G of the BA gauge is, like that of the older types, between 10 and 20 mbar^{-1}. Typical magnitudes are as follows:

$$\text{Filament emission to grid } (i_g) = 100 \ \mu\text{A}$$

$$G = 10 \ \text{mbar}^{-1}$$

$$\text{pressure} = 10^{-8} \ \text{mbar}.$$

Hence, the ion current (i_c) is 10^{-5} microamps. So, except for the higher pressures, an amplifier of the ion current is necessary.

The upper limit of the BA gauge is about 10^{-3} mbar, owing to loss of linearity and to the fact that the hot filament has in general a very short life at pressures greater than this. However thorium-coated iridium filaments have a longer life in 'gassy' conditions.

Poulter and Sutton (1981) have recently investigated the change with operating time of the sensitivity of ionisation gauges and report it to be of the order of 1% per 1000 h.

4.5.2 Other hot-cathode gauges

Ways of improving the performance of hot-cathode gauges include (*a*) using filaments with good electron emission at low filament temperature (see Turnbull *et al* 1962, p 89) thus reducing gas/filament chemical reactions, (*b*) employing a geometry of electric fields which increases the number of ions per emitted electron, hence both reducing the relative importance of the soft x-ray effect and requiring a lower filament temperature and so reducing the gas/filament reaction rate, (*c*) reducing the soft x-ray effect by arranging (Helmer and Hayward 1966) that the ion collector cannot 'see' any surfaces bombarded by electrons.

Chapter 6 contains further notes on gauges specially designed to minimise spurious residual collector currents, which are a major problem in measuring pressures below 10^{-10} mbar. However, because of its high sensitivity due to very long electron paths, achieved without use of a magnetic field (often inconvenient and sometimes unacceptable), the Orbitron gauge is mentioned here. It consists of a short thermionic filament located off-axis in a cylindrical positive ion collector, down the axis of which is a wire anode. The ion collector is grounded, the filament is about 10 V positive to it, and the central anode is a few hundred volts positive to the filament. The geometry is such that some of the electrons have an angular momentum with respect to the anode, and therefore describe long spiral paths round the anode before hitting it—the typical length is of the order of 1 m. In fact, they orbit the anode rather as a satellite orbits a planet. The electron current necessary to produce a given ion current and the soft x-ray effect are therefore very low and the sensitivity very high—values for sensitivity G of the order of 10^3 mbar being quoted (see Fitch and Thatcher 1968, Mourad *et al* 1964).

The problems particular to measuring very low pressures stem from the soft x-ray effect and other residual effects, which become dominant over the ionisation of gas molecules, rather than from difficulties in measuring the collector current itself. This will be further discussed (§ 6.2) in relation to ultra-high-vacuum techniques.

4.6 Cold-cathode Ionisation Gauges

Cold-cathode gauges are immune from a major hazard in hot-cathode gauges (i.e. filament burn-out caused by too high a pressure), but in most types it is necessary to provide a magnetic field, and hence a permanent magnet, to sustain the discharge at low pressures. Also, the voltages required are in general higher than those of thermionic filament gauges, and the unwanted pumping effect is greater.

4.6.1 *The Penning cold-cathode gauge*

In its simplest form, this consists (see figure 4.4) of a cylindrical anode at about $+2$ kV between two flat-disc cathodes at ground potential, placed between the poles of a permanent magnet which gives a field of about $\frac{1}{2}$ kG. As the pressure is reduced, a discharge is struck at about 10^{-2} mbar and the current, which is the sum of the positive ions reaching the cathodes and the secondary electrons liberated from it, decreases with pressure till it goes out at about 10^{-6} mbar. The electrons are constrained by the magnetic field to move in spiral paths in the potential trough between the two cathodes (see § AII.16), thereby increasing their free path before capture by the anode. Thus the chance of producing positive ions is enhanced and the discharge is maintained to much lower pressures than would otherwise be the case.

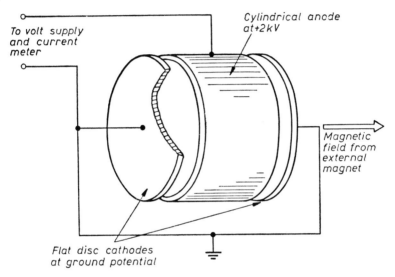

Figure 4.4   Elements of the Penning cold-cathode ionisation gauge.

The sensitivity is of the order of 10 A mbar^{-1} for N_2, so that the discharge current can be read on a microammeter. A typical two-range commercial instrument would cover 10^{-2}–10^{-4} mbar and 10^{-4}–10^{-6} mbar. The merit of the gauge is its ruggedness, but its disadvantages are that it is liable to sudden fluctuations of current and therefore of calibration (Leck 1964, p 115) and its cathodes are sputtered under the positive ion bombardment, causing pumping (which may be between 0.1 and 1.0 l s^{-1}) and impairing the insulation of the anode.

4.6.2   Other cold-cathode gauges

These devices, which are of magnetron geometry, have cylindrical symmetry and orthogonal electric and magnetic fields, the electric field being radial, the anode and the cathode coaxial cylinders, and the magnetic field axial. The magnetic field bends the paths of the electrons (see § AII.16, paragraph 1) so that they tend to return to the cathode whence they came unless they lose velocity by ionising a gas molecule and so decrease the restraining effect of the magnetic field. Further details of cold-cathode gauges designed specifically for measurement of very low pressures are given in chapter 6.

The Tesla coil, the high-frequency oscillations of which can excite electrodeless discharges in gases over a limited pressure range, cannot strictly be described as a cold-cathode gauge, but is useful as a qualitative pressure indicator over the pressure range 10 mbar to about 0.05 mbar. A typical instrument generates oscillations of 4.5 MHz frequency, and will, if held near the glass part of a vacuum system, produce luminous effects in the gas, the appearance of which is a very rough indication of pressure. It can be applied (hand-held) to any desired glass part of the system and immediately gives some information if the pressure is in its working range. If no discharge is seen the pressure is either many mbar or below 10^{-2} mbar. The colour of the discharge is an indication of the nature of the gas being excited (see Turnbull *et al* 1962, p 72).

4.7 The Knudsen Gauge

The Knudsen gauge may be classed as both mechanical and thermal, since it depends on the thermally produced momentum change when a cold molecule rebounds from a warmer surface, having been heated thereby. It can be designed for use at low pressures (e.g. 10^{-4}–10^{-7} mbar), has a calibration independent of gaseous species, (except in so far as they affect thermal accommodation (see § AII.6) on the warm surface) and does not need a hot filament, with its attendant problems of interaction with the gas. In some cases, these factors outweigh the disadvantage of its complexity and fragility. Steckelmacher (1951) gives an excellent account of this gauge, which will not be further described here.

4.8 The Viscosity Gauge

This gauge is of two types. In one—the oscillating vane type—a surface initially set in oscillatory motion is damped by molecules which impinge upon it, partake of some of its energy, and rebound. It is usable in the range 10^{-2}–10^{-5} mbar. It is claimed that, at a given temperature and pressure, the decrement of its oscillation is proportional only to the square root of the molecular weight of the gas, and hence the calibration, determined for one gas, is known for all.

The validity of this statement implies the questionable assumption that the degree of accommodation of energy and momentum on a surface is the same for all gaseous species. Mair (1967) has reviewed the consequences to the theory of oscillating gauges of various assumptions regarding the accommodation and angular distribution of rebounding molecules.

An outstanding advantage of the oscillating vane type is that it can be constructed entirely of fused silica, and hence can be used for work on corrosive gases and vapours which would attack the metal parts of other types of gauge. However, getting a reading takes time, which at 10^{-4} mbar may be of order 10^2 s.

The other type of viscosity gauge is the rotating type, in which a surface is kept in continuous rotation, its plane being parallel to a similar surface whose rotation is elastically constrained. Molecules strike the rotating surface and rebound, having acquired an additional component of momentum parallel to the direction of movement. Part of this is transferred to the elastically constrained member, causing it to rotate to an equilibrium position which is a function of pressure. The gauge is capable of measuring pressures as low as 10^{-6}–10^{-7} mbar. An account of the theory and practice of viscosity gauges is given in Leck (1964, pp 35 ff) and Dushman (1962, pp 224 ff), but as these gauges are not widely used they will not be further discussed here.

4.9 Gauge Calibration

The three methods commonly used are (*a*) comparison with a McLeod gauge, (*b*) the reduction, assuming Boyle's Law, of the pressure of a known small volume of gas at a known (e.g. atmospheric) pressure by a known expansion and (*c*) a dynamic method, depending on the pressure drop across an aperture of known conductance. Meinke and Reich (1967) have compared methods (*b*) and (*c*) for He and Ar over the range 10^{-2}–10^{-6} mbar.

4.9.1 *Comparison with a McLeod gauge*

This gauge, already described in § 4.3 can, if proper precautions are taken, be used as a standard to calibrate BA ionisation gauges down to pressures approaching 10^{-5} mbar (see Elliott *et al* 1967), with an accuracy of 1 or 2%, but only for non-condensable gases. Since it employs mercury, an associated cold trap with its consequent correction for the vapour-stream effect (see § 4.3.1) is necessary. It is, however, widely used as a calibrating gauge and its sources of error and the precautions to minimise them have been extensively studied.

4.9.2 Gas-expansion method

In this technique (originally due to Knudsen) a small known volume V_1 of calibrating gas at a known pressure is expanded into a much larger known volume V_2, a small portion of which, V_3, can be isolated. The remainder of the gas is pumped away and the gas in V_3 again expanded into a large volume which may conveniently be V_2. The process is continued as necessary and the final pressure calculated by Boyle's law. Two possible sources of error are (*a*) the outgassing of the apparatus which would add unknown amounts of gas, and therefore pressure to the apparatus and (*b*) adsorption causing disappearance of gas. Effect (*a*) can be minimised by suitable pre-treatment, but effect (*b*) is in principle unavoidable, though in practice it is apparently not significant with permanent gases such as N_2 and Ne (see Elliott *et al* 1967).

4.9.3 The dynamic method

This is a steady-state method, in which a known and constant influx of gas $\dot{Q}$, conveniently measured at or near atmospheric pressure, enters the upper part of a chamber divided into two compartments by a diaphragm pierced by an aperture (see the schematic figure 4.5) of conductance F, calculated from its area†, the temperature and the molecular weight of the gas. The lower part of the chamber is connected to a diffusion pump of such speed that $P_1 \gg P_2$, P_1 and P_2 being the pressures above and below the aperture. The gauge to be calibrated is connected to the upper compartment at pressure P_1.

In the steady state

$$\dot{Q} = F(P_1 - P_2)$$

---

† The correction to F, due to the finite ratio of the aperture area to the diaphragm area is discussed by Bureau *et al* (1952). F itself is calculated from equation (2.11).

or

$$P_1 = \frac{\dot{Q}}{F}\left(\frac{1}{1 - P_2/P_1}\right). \tag{4.7}$$

P_1/P_2 is obtained from a gauge G which can be connected, by valves A and B, to either side of the aperture at will. Since $P_2/P_1 \ll 1$ it need not be determined with great accuracy, and G need not be calibrated in absolute terms since it only furnishes a ratio. In the steady state, there are no adsorption errors, but it is necessary that outgassing should be small compared with $\dot{Q}$. For details Leck (1964, pp 168 ff) may be consulted.

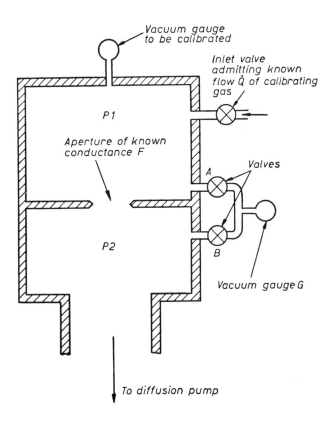

Figure 4.5 Dynamic calibration of vacuum gauge (schematic).

4.10   Thermal Transpiration and Pressure Measurement

Consider two chambers 1 and 2 at temperatures T_1 and T_2, separated by a small orifice at which the temperature discontinuity occurs. If the pressures on the two sides are both so low that gas–gas collisions are negligible compared with gas–wall collisions, the molecules going from 1 to 2 will have a mean velocity $\bar{c}_1$, appropriate to T_1, and those from 2 to 1 will have a mean velocity $\bar{c}_2$ appropriate to T_2. In the steady state the net flux through the orifice between the chambers must be zero. Hence (see § AI.6)

$$n_1\bar{c}_1 = n_2\bar{c}_2. \tag{4.8}$$

Thus, from equation (AI.3)

$$\frac{n_1}{n_2} = \left(\frac{T_2}{T_1}\right)^{1/2}. \tag{4.9}$$

But (see equation (AI.5))

$$P = nkT$$

$$\therefore \frac{P_1}{P_2} = \frac{n_1 T_1}{n_2 T_2}$$

$$= \left(\frac{T_1}{T_2}\right)^{1/2}. \tag{4.10}$$

On the other hand, at high pressures, where gas–gas collisions are predominant and continuous fluid mechanics apply, to a good approximation

$$P_1 = P_2. \tag{4.11}$$

So for the case of an aperture, there is, as P increases, a continuous transition in behaviour between the regimes corresponding to equations (4.10) and (4.11). Note that equation (4.10) is valid for an *aperture* connecting two vessels at different temperatures, but not necessarily for a *tube*, which is the situation when a pressure gauge at room temperature is connected by tubulation to a work chamber at a different temperature. Edmonds and Hobson (1965), working with He and Ne, found equation (4.10) valid at low pressures for apertures, but not, in general, for pyrex tubes, for which it overestimates P_1/P_2 for $T_1/T_2 > 1$. They also give many references to earlier work and conclude that there is no rigorous theoretical or experimental basis for equation (4.10) as a general limit, except for apertures. The matter is reviewed by Redhead *et al* (1968, pp 281 ff). See also Hobson (1969), who reports that for a pyrex tube leached in HCl, equation (4.10) is valid at low pressures. He concludes that leaching produces an atomically rough surface causing all molecular scattering to be cosine, resulting in the ideal law for transpiration.

4.11 The Determination of Partial Pressures

The pressure gauges described in previous sections give information on the total pressure of the gas phase, but none on its composition, which however is often useful (e.g. to identify the source of virtual or real leaks) and is crucial in some experiments, such as the determination of low-pressure chemical kinetics.

4.11.1 *Partial pressures by mass spectrometry*

In nearly all types of mass spectrometer, the gas is ionised by means of an electron current from a hot filament, and the relative abundance of the resulting ions determined, separated according to the value of their mass/charge ratio.

As the ionisation probability is a function of the molecular species, the relative abundance of the ions is not the same as that of the neutral molecules from which they are derived. Also, the instruments do not present a spectrum of m alone. This can cause ambiguity since, for example, singly ionised N and doubly ionised CO both give peak signals at $m/e = 14$.

In all types of mass spectrometer, the presentation of an ordered spectrum of m/e requires the motion of the ions to be controlled by the parameters of the instrument, undisturbed by chance collisions with gas molecules. Since the trajectory of the ions in a mass spectrometer is usually a few centimetres and their mean free path is approximately equal to that of neutral molecules, there is an upper limit to the pressure of operation, generally of the order of 10^{-4} mbar. (See § AI.9).

4.11.2 *Important characteristics of mass spectrometers*

A mass spectrometer should have a large and linear response for a given ion, be able to scan the whole mass spectrum quickly, be bakeable (to avoid 'memory' of previous use), have little adsorption of gases to be analysed (which means minimum wall area), and little chemical reaction with the analysed gases, which demands minimum area and temperatures of the thermionic filament. The ability to resolve neighbouring *lines* in the mass spectrum (called the mass resolution) is also of great importance.

The mass resolution R is defined in several different ways. A common definition is

$$R = M/\Delta M \qquad (4.12)$$

where M is the mass number† and $\frac{1}{2}\Delta M$ is the deviation on each side of M at which the intensity of the signal due to this mass number has fallen to $f\%$ of its peak value, where f is often either 0.1 or 1%.

Suppose that an instrument is said to have $R = 30$, using the 1% definition. This means, for example, that the signal due to $M = 45$ would fall to 1% of its peak value at a distance of $\frac{3}{4}$ of a mass unit on either side of the peak value. Another often used definition of R is the mass number at which the height of the bottom of the valley between two peaks of equal height and one mass number apart is ten per cent of the height of either peak. In assessing the resolution of an instrument it is important to know which definition of R is being used. For a comparison of the various definitions, see Dennis and Heppel (1968).

4.11.3   Types of mass spectrometer

Mass spectrometers may be classified as static or dynamic. In the former type (in order to produce a peak signal for a particular value of m/e) the parameters of the instrument are held constant. In the dynamic instrument, one or more parameters are varied at a frequency which allows ions of a particular value of m/e (unique to that frequency) to be collected.

A popular mass spectrometer‡ of the static type is the magnetic deflection instrument, in which ions produced by impinging electrons (as in a hot-cathode ionisation gauge) are electrostatically extracted and

---

† M is the molecular weight of an ion divided by the number of electrons it has lost; e.g. M for doubly ionised CH_4 is 8.

‡ In cases where the presence of a magnetic field is compatible with the other requirements of the situation.

caused to move in a circular arc of fixed radius R by a fixed magnetic field H (see § AII.16) at the termination of which is an ion collector. To each value of m/e there corresponds an extraction voltage V, (and therefore an initial velocity perpendicular to the fixed H) which will cause the ions to move in an arc of radius R, and thus reach the ion collector. The m/e spectrum is therefore scanned by varying V and observing the resultant ion current reaching the ion collector.

Another type of static instrument is the omegatron (see e.g. Robinson 1968, pp 111 ff), in which ions are produced by an electron beam parallel to a constant magnetic field H under the influence of which they orbit in planes perpendicular to H with a frequency ω, which is determined only (see § AII.16) by their e/m. However, the orbital radius R is proportional to the square root of their kinetic energy. An RF electric field of variable frequency is applied in the plane of the orbits (i.e. perpendicular to H) and the ions can gain energy from this when the RF frequency becomes equal to ω. Thus, for each value of e/m there is an RF which will cause the path of the orbiting ions to spiral outwards until the ions impinge on a suitably placed collector. The current from this is amplified and monitored on an external recorder, so that the mass spectrum may be displayed as a plot of signal strength against radio frequency.

A type of dynamic instrument which does not require a magnetic field is the quadrupole mass spectrometer. In this, the ions are extracted from the ion source, and move down the longitudinal axis of an array of four cylindrical electrodes symmetrically placed with their centres $4r_0$ apart, where r_0 is the radius of any one. Two diametrically opposite cylinders at a voltage U with respect to the ion source are connected together and the other two are also connected and have a common potential of $U + V \sin \omega t$. The frequency ω is a constant of the instrument, with the result that, for given values of U and V, only ions of a specific value of m/e can pass down the axis of the array and reach the collector. For other values of m/e they execute oscillations of increasing amplitude and are collected on the electrodes. Hence a scan of m/e is achieved by varying V and U, but the ratio V/U must be kept constant, since the resolution is determined by this. In a variant of this system, only two electrodes—a cylinder within a channel of rectangular cross section—are used (see Robinson 1968, p 126).

In the time-of-flight mass spectrometer, a pulse of ions is produced and accelerated by a controlled voltage (and therefore to a known velocity) and allowed to travel at this velocity down a tube of known length to an ion collector. If the ions are extracted from the ion source by a potential V and travel down a tube of length L at a velocity v to the collector,

$$eV = \tfrac{1}{2} mv^2 \tag{4.13}$$

and t, the time of flight is given by

$$t = L \left(\frac{m}{2\,eV} \right)^{1/2}. \tag{4.14}$$

Hence ions of various m/e are formed into bunches and arrive at the collector after various time intervals, which can be displayed on a cathode ray oscillograph with a fast time sweep (see Bailey *et al* 1971, pp 461–4). One major advantage of this instrument is its very high rate of scan. Other types of mass spectrometer include the Favitron and the linear accelerator type (see Robinson 1968, pp 127, 128).

For reviews of mass spectrometers, see Steckelmacher (1972), Leck (1964, ch 7) and Coleman (1967). Coleman reviews the operating principles of five basic types and analyses in detail the specification and performance of magnetic deflection instruments and quadrupole or monopole types.

See also Colligon (1982) which is a review by a number of authors of the present-day (1982) use of mass spectrometers for industrial control.

4.11.4   *Measurement of partial pressures by sorption methods*

The desorption rate of a gas from a surface is dominated by the Boltzmann factor, $\exp(-q_d/kT)$, q_d and k being the desorption energy and gas constant per molecule (see § AII.19). Suppose a surface located in a vessel containing a partial pressure of a gas is cooled to a temperature T such that $q_d \gg kT$. Adsorption of the gas occurs, and unless chemical combination also occurs, this will be reversed on re-heating. Because the temperature coefficient of desorption is so large, owing to the Boltzmann factor, the desorption (which may be measured by a total pressure gauge in the system) becomes appreciable rather suddenly as the surface temperature rises, at a temperature T_d determined by q_d, and hence by the particular gas desorbed. This is the basis of a simple but not very precise method of measuring partial pressures.

The temperature T_d is shown as a point of inflection in the pressure/temperature curve if the analysis is done in a closed system or, if the system is continually pumped, as a pressure *peak*, the height of which, when the initial coverage was less than a monolayer, is a measure of the original partial pressure of the gas in the system. The value of T_d does not depend greatly on the rate of heating up of the surface, so that its value and the height of the peak allow both the adsorbed species and the partial pressure which give rise to it to be deduced.

The adsorption/desorption surface, whose temperature is caused to increase with time in a predetermined manner, is often a W filament. A typical procedure for determining partial pressures in a low-pressure vessel ($P \approx 10^{-10}$ mbar) is first to 'flash' the W to about 2500 K to remove all adsorbed species. It is then allowed to cool to 300 K and, after a known period (t_c) at this temperature (to allow adsorption), is raised in a time of order 10^2 seconds to its original high temperature, the pressure and temperature being recorded. The amount of gas desorbed at T_d is a measure of the gas sorbed during t_c, while T_d itself determines the species. Prior calibration of the apparatus may be done by the insertion of known amounts of known gases or with a mass spectrometer. For details Redhead *et al* (1968, pp 355 ff) should be consulted.

At relatively high pressures, say 10^{-2} mbar, a crude analysis for H_2O and CO_2 is possible by cooling a Pyrex side tube of the vacuum vessel whose contents it is desired to analyse, and noting the temperature of the Pyrex surface at which rather sudden decreases of pressure occur. For example, H_2O is virtually completely condensed (i.e. removed from the gas phase) at temperatures between $-40\,°C$ and $-100\,°C$ and CO_2 between $-140\,°C$ and $-180\,°C$. In addition the amounts of any H_2 and O_2 present can be estimated by using heated Pd or Ag tubes to admit further quantities of H_2 or O_2 as necessary, promoting the $2H_2 + O_2 = 2H_2O$ reaction on a heated Pt filament, and determining the amount of water produced (see Carpenter and Mair 1953). The above methods are laborious, limited in scope to those gases which are adsorbed at the lowest temperature available, and of low precision, but can be useful when convenience or expense preclude the use of a conventional mass spectrometer.

References

Allsopp H J, Bradshaw F J, Stanford R H and Wadsworth N J 1969
    Vacuum **19** 61–7
Bailey C A, Kilvington J and Robinson N W 1971 *Vacuum* **21** 461–4
Bureau A J, Laslett L J and Keller J M 1952 *Rev. Sci. Instrum.* **23** 683

Carpenter L G and Mair W N 1953 *RAE Tech. Note* No Met 174

Cleaver J S 1967 *J. Sci. Instrum.* **44** 969

Cobic B, Carter G and Leck J H 1961 *Vacuum* **11** 247–57

Coleman J W 1967 *Technical Memorandum No 110152* of the BP Research Centre, Sunbury-on-Thames, Middlesex, England

Colligon J S ed 1982 *Vacuum* **32** 157–84

Dennis N T M and Heppel T A 1968 *Vacuum System Design* (London: Chapman and Hall) pp 97 ff

Dushman S 1962 *Scientific Foundations of Vacuum Technique* ed J M Lafferty (New York: Wiley)

Edmonds T and Hobson J P 1965 *J. Vac. Sci. Technol.* **2** 182–97

Elliott K W T, Woodman D M and Dadson R S 1967 *Vacuum* **17** 439–44

Fitch R K and Thatcher W J 1968 *J. Sci. Instrum.* **45** 317–9

Green M and Lee M J 1966 *J. Sci. Instrum.* **43** 948–9

Helmer J C and Hayward W H 1966 *Rev. Sci. Instrum.* **37** 1652–4

Hobson J P 1969 *J. Vac. Sci. Technol.* **6** 257–9

Leck J H 1964 *Pressure Measurement in Vacuum Systems* 2nd edn (London: Chapman and Hall)

Lewin G 1965 *Fundamentals of Vacuum Science and Technology* (New York: McGraw-Hill) p 96

Mair W N 1967 *RAE Tech. Rep.* 67024

Meinke C and Reich G 1967 *J. Vac. Sci. Technol.* **4** 356–63

Mourad W G, Paulty T and Herb R G 1964 *Rev. Sci. Instrum.* **35** 661–5

Poulter K F and Sutton C M 1981 *Vacuum* **31** 147–50

Redhead P A, Hobson J P and Kornelsen E V 1968 *The Physical Basis of Ultra-High Vacuum* (London: Chapman and Hall)

Roberts J K and Miller A R 1955 *Heat and Thermodynamics* 4th edn (Glasgow: Blackie and Son) ch II §6

Robinson N W 1968 *The Physical Principles of Ultra-High Vacuum Systems and Equipment* (London: Chapman and Hall)

Rothe E W 1964 *J. Vac. Sci. Technol* **1** 66–8

Steckelmacher W 1951 *Vacuum* **1** 266–82

——1965 *J. Sci. Instrum.* **42** 63–76

——1972 *Laboratory Equipment Digest* No 12

Tunnicliffe R J and Rees J A 1967 *Vacuum* **17** 457–9

Turnbull A R, Barton R S and Riviere J C 1962 *An Introduction to Vacuum Technique* (London: George Newnes) p 189

5 Materials and the Design of Apparatus

5.1 Material Properties and Design Requirements

Space does not permit detailed description of the properties of the many materials used in vacuum technology, for which reference may be made to existing authoritative sources, for example, Rosebury (1965) and Espe (1966–8). This chapter will indicate some considerations governing material choice, briefly mention a few common materials and vacuum accessories (such as valves), and outline a typical design procedure. Some notes on the troublesome and time-wasting matter of leak detection are included.

The choice of materials from which a system is constructed is strongly influenced by its intended use. For example, if the system is an industrial one, and ruggedness is desired, metal is usually preferable to glass. In research laboratories, glass may be suitable, but the choice may be influenced by the relative availability of glass- and metal-working facilities. In a metal system, ease of leak-tight welding or brazing must be considered. The operating temperature is important; if part of the apparatus is at a low temperature, neither elastomers which become brittle nor stressed parts of metal liable to brittle fracture can be used. A low thermal conductivity may be desirable to limit heat conduction from the warm to the cold parts of the apparatus (and hence to conserve refrigerant), or a high one to improve isothermality. At high temperatures, vapour pressure, which is a strong function of temperature (see § AII.3), precludes the use of some metals (e.g. zinc distils out of brass) and thermal degradation precludes the use of organic substances. The outgassing rate (see § 5.8) is always an important property in high-vacuum apparatus, since it limits both the final pressure attainable and the time taken to reach it.

5.2 Glass and Glass-to-Metal Seals

Borosilicate (so-called 'hard') glasses are generally used in preference to 'soft' ones, because they have a low coefficient of linear expansion (usually between 3 and $6 \times 10^{-6}\,°C^{-1}$) and therefore do not easily crack when subjected to a thermal gradient, and a relatively high (up to about 450°C) service temperature. However, soft glasses are often used in large-scale production, e.g. soda-lime glasses as the envelopes for small lamps and thermionic valves.

The most common hard glass used in vacuum work is Pyrex (Corning 7740) and many vacuum gauges are now made in glasses which seal directly to it. Otherwise, a graded seal (consisting of several glasses of gradually decreasing expansion coefficient between that of the gauge and the Pyrex of the vacuum apparatus to which it is connected) is used. Pyrex apparatus can be cooled with LN_2 (77 K) without much risk of cracking and has an upper service temperature of about 450°C, but Bills and Evett (1959) warn that it may be a disturbing factor in certain *clean-surface* scientific experiments, as it appears to decompose at temperatures above about 350°C, liberating H_2O and (more seriously) certain contaminates not readily pumped out of vacuum systems. Donaldson (1962) also calls attention to certain undesirable effects. However, for work which does not call for ultra-cleanness, Pyrex remains a most useful material.

Fused silica (i.e. amorphous glass derived from crystalline quartz) has a service temperature up to about 1000°C, because of its low coefficient of linear expansion ($\sim 0.5 \times 10^{-6}\,°C^{-1}$), is virtually unbreakable by thermal gradient, and is a much better electrical insulator than Pyrex. It can be sealed to the latter via a number of intermediate glasses of carefully graded expansion coefficients.

The uses of glass/metal seals in vacuum systems include the following:

(*a*) to make a junction in the vacuum envelope between the two materials,
(*b*) to introduce electrical leads into glass vacuum envelopes,
(*c*) to introduce electrical leads into metal vacuum envelopes, the glass serving as an insulator.

Purpose (*c*) can also be served by insulators other than glass (see e.g. Kohl 1964).

In making glass/metal seals it is necessary that the glass should bond to the metal or its oxide, and that either their expansion coefficients should not be greatly different or the metal should, by elastic or plastic deformation, conform to the dimensions of the glass as its temperature changes. An example of approximately matched expansion is the

W/Pyrex seal, often used for introducing electrical leads into vacuum systems. Another example of this is the Kovar/Kodial seal—the former being an alloy of Ni, Co and Fe which matches Kodial glass, which itself can be sealed to Pyrex via one intermediary glass—the *graded seal* technique.

The alternative *metal-conforming* method is exemplified by the House-keeper seal, in which a feather-edge of a copper tube is sealed into a Pyrex tube. A similar technique has been used for joining Pyrex and stainless steel. Making metal/glass seals is a highly skilled technical job. Fortunately, most types are commercially available. For an excellent review of possibilities, see the book by Green (1968, ch 3) on constructing small vacuum systems. Holland's book on the properties of glass surfaces (Holland 1961) contains many data useful in vacuum technology.

An interesting development in insulating materials for use *in vacuo* is Corning code 9658 machinable glass ceramic. This is a devitrified glass, i.e. a glass which by the addition of nucleating substances and heat treatment has changed into a mass of very small crystals. Code 9658 is marketed under the name of Macor†. An outstanding characteristic, stemming from its very fine grain size, is its machinability by standard metal-working tools to precision tolerances. Its mechanical and thermal properties are summarised by Mog (1976).

The vacuum properties of Macor are described by Altemose and Kacjon (1979) who state that up to 600°C it can be treated like glass vacuum system components while its helium permeability is an order of magnitude less than that of code 7740, i.e. of Pyrex.

5.3 Metals

The most commonly used metals for vacuum envelopes are brass, copper and stainless steel. Brass, being an alloy of Cu and Zn, cannot be used above 100–150°C because at high temperatures *in vacuo* the Zn distils out‡. Copper, having a much lower vapour pressure, can be used to much higher temperatures *in vacuo*, but if used as part of a vacuum envelope, its outer surface, exposed to air, will be seriously oxidised above a few hundred degrees centigrade. It is particularly useful when heat has to be conducted to or from certain parts of the apparatus. Oxygen-free high-conductivity copper (OFHC) should generally be used,

---

† Obtainable in the UK from William McGeoch and Co (Birmingham) Ltd.
‡ Similarly, high vapour pressure precludes the use of hard solders containing Cd.

since copper containing oxygen reacts at temperature with H_2 to produce H_2O and some grades of ordinary copper contain impurities which very seriously reduce its thermal conductivity, e.g. 0.4% of arsenic, often deliberately added to improve mechanical properties, may reduce both electrical and thermal conductivities to less than half those of pure copper.

A suitable steel for welded vacuum vessels is BS 970 EN 58G (American equivalent AISI 347). This is a non-magnetic stainless steel of the 18/8/1 type, the latter component being Nb, whose purpose is to combine with any C present and prevent formation of chromium carbide, which by depleting the Cr content of the grain boundaries leads to weld decay. Where conditions are not so stringent as regards freedom from small leaks, and minimum magnetic permeability is not a requirement, BS 970 EN 58B (the American equivalent AISI 321, in which Ti is substituted for Nb and the Ni content is lower) may be used.

Mild steel is sometimes used for vacuum envelopes, but because it rusts when let down to atmosphere, thereby increasing its outgassing rate in subsequent service, it is usually nickel or chromium plated. Also, at low temperatures ordinary mild steel should be used with caution in vacuum vessels owing to the risk of brittle fracture caused by 'locked up' stresses. Turnbull *et al* (1962, ch 5) give a useful list of the physical properties of metals of interest in vacuum work.

5.4 Plastics

Plastics, in the sense of rigid organic materials (as opposed to elastomers, dealt with in § 5.6), are little used as materials of construction in high vacuum technology. It is suggested, however (Laurenson, private communication), that a polyamide gasket clamped between stainless steel flanges constitutes a satisfactory seal which will stand baking and cycling to 300 °C.

Minor uses of plastics include PVC tubing for vacuum pipe lines at pressures greater than 10^{-3} mbar (Hartman 1967), and synthetic resin adhesives as vacuum-tight cements in experimental vacuum apparatus (e.g. Sayers 1960). Barton and Govier (1965) have studied the outgassing of certain plastics, including Araldites, Mycalex, Perspex and PTFE, analysing the evolved species with a mass spectrometer. For useful information on epoxy resins (and on AgCl used as a high-temperature cement) see Green (1968, pp 77–8). Kendall and Zabielski (1966) have published data on high-temperature ($\sim$300 °C) insulating adhesives for vacuum applications.

5.5 Waxes and Greases

Waxes are useful for making vacuum seals in semi-permanent experimental apparatus. They should have a low vapour pressure and a softening point above the highest likely service temperature, and should not decompose when heated during application. The Shell Chemical Co Ltd market a range of vacuum waxes with melting points from 45–85 °C and vapour pressures, *after evolution of dissolved air*, of 10^{-3} mbar at 180 °C.

Greases are used as sealants and lubricants in glass vacuum taps and ground joints. Again, a number of greases are available with vapour pressures at 20 °C ranging from 10^{-3}–10^{-10} mbar after the evolution of dissolved air. This qualification is important, since the greases in a newly assembled vacuum system are often sources of considerable gassing—a fact demonstrated by the rise in reading of a pressure gauge when a newly greased tap is turned. According to Allsopp (1961), vapours from vacuum waxes and greases may be 'cracked' on a hot Pt filament, producing significant quantities of low molecular weight gases.

The elastomer O-rings of demountable couplings are often lightly greased to improve leak tightness. After so doing, care should be taken not to allow the greased O-rings to pick up dust particles or fibres which may spoil their seating and degrade rather than improve their vacuum integrity.

5.6 Demountable Couplings and Motion *in Vacuo*

Demountable couplings are usually circular in cross section, and their vacuum tightness depends upon the deformation of an elastomer or metal gasket (irreversible in the latter) compressed by some mechanical device. The gaskets are either elastomers, used in systems not subjected to severe baking, or metals, used in UHV systems baked to temperatures up to about 450 °C.

In designing an apparatus or setting up a vacuum laboratory it is well to study carefully the various types of coupling commercially available, and having chosen one, to standardise on that. The unnecessary use of several types can lead to much expense and loss of time.

The subject of couplings, though mundane, is very important in practice. An attempt to describe the various options would be out of place in a small book, and superficial treatment useless. But a few broad points with reference to sources of detailed information are given below.

The important qualities in elastomers are maximum service temperature, impermeability to gases, outgassing rates, resistance to oil, and mechanical properties, especially the degree of resistance to permanent 'set' under compression.

A common elastomer used in vacuum seals is nitrile rubber (i.e. a copolymer of butadiene and acrylonitrile) with a maximum service temperature of about 80 °C. For temperatures up to 200 °C, a fluorinated elastomer can be used, but as this tends to acquire a permanent set it should preferably be confined in a groove or recess of cross section area not more than 5% greater than that of the gasket itself. For an account of elastomer seals and associated couplings based on extensive first-hand experience, see Turnbull *et al* (1962, ch 5).

O-ring gaskets (i.e. gaskets with a circular cross section) usually have a leak rate less than 10^{-5} mbar l s^{-1}, in e.g. pipe couplings with a diameter of a few centimetres. There are numerous elastomer-type couplings. A good system should include a range of couplings of various diameters, with associated valves, elbow bends and T-junctions, and adaptors for changing from one diameter of tube to another.

Metal demountable seals depend upon the plastic irreversible flow of metal gaskets, compressed by forces exerted on metal flanges by bolts. The metals should have low vapour pressures, high melting points, and reasonable resistance to oxidation.

In the Varian Conflat seal, an originally flat copper ring gasket is compressed between projections in two similar stainless steel flanges, the amount of plastic flow and therefore stress relief in the copper being limited by the geometry of the flanges, which are bolted up until they come in contact.

In another type, a copper gasket of diamond cross section is compressed between two stainless steel flanges, in one of which there is a circular groove, whose depth is 85% of the apex-to-apex dimension of the gasket. The mating surfaces of the flange make contact when bolted up. In a simpler but reputedly less satisfactory variant of this type, the faces of the two flanges are identical, both being plane.

A simple seal consists of a gold wire gasket compressed between bolted flanges to about half its original diameter. In all the above seals, the parts of the flanges in contact with the deformable gasket should have a fine machine finish and, in particular, should be free from *radial* scratches.

The seals mentioned above are merely a selection from a considerable number of types, about the relative merits of which there is no universal agreement. A worker considering the use of bakeable metal seals should consult the work of authorities such as Robinson (1968), Turnbull *et al* (1962, ch 5) and Lewin (1965).

In vacuum systems not subject to baking, transmission of rotational motion from outside may be by means of greased rotatable cone-and-socket joints or by a highly polished metal shaft passing through a greased and suitably supported elastomer.

For baked systems, a circular pendulous motion of a shaft may be transmitted by a metal bellows, usually of stainless steel. By moving the outer (atmospheric) end, the inner (vacuum) end of the shaft can be manipulated so as to cause the rotation of a bearing itself wholly within the vacuum. For details of this and other like devices, see Turnbull *et al* (1962, pp 127 ff) and Green (1968, pp 128 ff).

A recent development is a rotating shaft seal for introducing high-speed rotation into a vacuum chamber. The sealant is a magnetic fluid held in place by a static magnetic circuit. The magnetic fluid consists of a suspension of very small (≈ 100 Å) ferromagnetic particles in a low vapour pressure oil. For details see Raj and Grayson (1981).

5.7 Vacuum Taps and Valves

For small laboratory apparatus, greased pyrex stopcocks are still often used. They consist of a conical glass key carefully ground into an outer glass barrel, to which are fused tubular glass side arms. In one form the key is traversed by a diametral bore, the tap being open when the bore is in line with side arms. In use, continued rotation of the tap tends to produce circular streaks in the grease, which may eventually lead to partial short-circuiting of the tap, even when it is nominally closed, with the axis of the bore perpendicular to side arms. This particular type of leak is obviated in a second form in which the bore passes obliquely across the key, the two side arms being offset.

Glass stopcocks with bores of more than about 2.5 cm are unwieldy and often become difficult to turn as the grease ages. By means of glass/metal tubular seals (see § 5.2), it is possible to incorporate glass stopcocks in otherwise all-metal systems, and this is sometimes advantageous, since glass stopcocks greased with a low vapour pressure grease probably outgas less than metal valves of the elastomer diaphragm type.

In the latter, the seal between the two sides of the valve is made by an elastomer diaphragm moved into the closure position by an externally operated screw mechanism. In a variant of this, the closure is effected by the movement, perpendicular to its own plane, of a circular plate carrying an O-ring in a groove on its perimeter, isolation from the external atmosphere being by means of bellows.

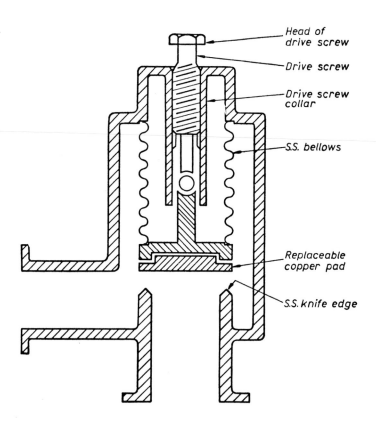

Figure 5.1   Bakeable all-metal valve (schematic).

Other types of non-bakeable valve include the gate type, in which closure between the two sides is made by the movement of a horizontal plate in its own plane, the seal being an elastomer, and the quarter swing (butterfly) type, in which a disc is caused to rotate about a horizontal shaft crossing the diameter of the circular aperture to be closed.

Both gate and butterfly types are often used immediately above the mouth of a diffusion pump or above the water-cooled baffles (see § 3.3.1) if any are fitted. A third type, commonly used for the same purpose, is the plate valve, in which a horizontal circular plate of metal is located immediately above the mouth of the diffusion pump and can be moved vertically by means of a lever mechanism operated from outside. In the closed position, the plate rests on the circular mouth of the diffusion pump, the vacuum seal being provided by the peripheral O-ring; in the open position, the plate is suspended a centimetre or so above the pump mouth, thus itself forming a baffle which tends to prevent oil vapour from the diffusion pump from going unimpeded to the work chamber, but allowing pumping to proceed without undue reduction of speed.

Valves in apparatus to be degassed by baking at above 200 °C must be of all-metal construction. Figure 5.1 shows the principle of one common type, in which vacuum-tight closure is obtained by forcing a pad of a soft metal, such as copper, onto the mouth of a sharp-edged orifice of a hard metal, such as stainless steel. The driving mechanism, usually a threaded shaft, is at atmospheric pressure and is separated from the evacuated space by a metal bellows.

5.8 Outgassing

When a solid is first exposed to a vacuum, gas is desorbed and the desorption continues indefinitely, but at a decreasing rate. The released gas may in principle be considered as consisting of two kinds, that desorbed from the surface and that which diffuses from the interior, but since the surface gas is constantly replenished by that coming from the interior, the distinction is somewhat academic. However, it is often true that the initial surface-desorbed gas is mainly H_2O, at any rate for metals and glasses. There exists a considerable body of data on the rate of outgassing of various materials as a function of time. The rate, expressed in $mbar \, l \, s^{-1} \, cm^{-2}$, decreases in a roughly exponential manner with time. For example, a typical elastomer O-ring might have an initial rate of

10^{-6} (in the above units) decreasing by a factor of 2 at the end of 5 h and another factor of 2 at the end of 10 h. Corresponding figures for slightly rusty mild steel might be 10^{-6} when first exposed to vacuum, with a decrease by a factor of about 10 in the first 5 h and by a similar factor in the next 5 h. Stainless steel of the 18/8/1 type might have an initial rate about a decade lower, but the rate decays in about the same way as for mild steel.

Outgassing/time curves for various constructional materials are given by Power (1966, pp 392 ff) and by Dennis and Heppell (1968, pp 126 ff), who themselves give references to further data. The subject is also reviewed by Holland *et al* (1974) and Elsey (1975). The latter gives methods of measuring and reducing outgassing and quotes rates for various materials.

The outgassing rate of a given material is strongly dependent on its previous surface treatment. For instance, the rate after 1 h for 'raw' stainless steel is diminished by a factor of 5 by electropolishing (Dennis and Heppel 1968). Also, outgassing increases rapidly with temperature, so that even a mild bake (degassing) at 100 °C greatly reduces the rate when the material returns to room temperature. If the system has to be 'let up' periodically to atmosphere and then resealed and repumped, the pump-down time can be significantly reduced by letting up with dry N_2 instead of with wet atmospheric air.

5.9 Design Procedure

In a design it is necessary to consider the following properties:

the physical size of the work chamber;
the temperature of the work chamber;
the electrical connection to the inside of the work chamber;
the mechanical motions to the inside of the work chamber;
the degree of vacuum required;
the required precision of pressure measurement;
the pump-down time to the required vacuum; and
the skill of the intended operator.

If a work chamber of the bell jar type (i.e. a bell jar mounted on a flat base plate, an L-section circular elastomer being interposed as a seal) is contemplated, the maximum diameter should not exceed 30 cm if the bell jar is glass†; above this diameter the bell jar will be metal—usually stainless steel. If the power input to the contents is more than a few tens of watts, the base plate (and possibly the bell jar) is usually water-cooled.

If the vacuum envelope is of the demountable gasket type, e.g. bell jar or 'top hat' on a base plate, the electrical and mechanical connections are conveniently brought in through a 'service' collar, which is interposed between the base plate and the main vacuum envelope. The advantage of this arrangement is that, if a modification to electrical or mechanical inputs is required, it can be effected by modifying the collar without interfering with the base plate, which is usually connected to the pump via a valve and (often) by a LN_2-cooled vapour trap.

For pressure measurement, a Pirani gauge is useful to cover the range before it is safe (i.e. 10^{-3} mbar or preferably less) to switch on a hot-cathode ionisation gauge. If only a rough indication in the range 10^{-2}–10^{-6} mbar is required, then the Penning gauge, having a cold-cathode and therefore immune from burn-out, has much to commend it.

If the apparatus is to be operated by unskilled personnel, safety devices to guard against accidents, such as the switching on of diffusion pumps without backing pumps or omitting to turn on cooling water, are desirable. If the required pressure in the work chamber is comparatively high, say not much less than 10^{-1} mbar, the pump-down time of the empty chamber may be calculated by the procedure of § 2.8. If, however, pressures of 10^{-3} mbar and below are required, virtual leaks (i.e. desorption) become important and a different method must be used.

The following numerical example is given merely to illustrate *method*, using figures which may give some 'feel' for orders of magnitude.

Suppose that the problem is to choose pumping equipment for a small 'top hat' type stainless steel vertical cylindrical chamber of height 60 cm and diameter 30 cm. The service collar carrying electrical leads and other services is sandwiched between two elastomer gaskets. Above the collar is the bottom flange of the top hat, and below it the base plate of the apparatus which is connected to a diffusion pump via a baffle valve. Assume, as is likely, that a combination of an oil diffusion pump and a mechanical backing pump has been chosen. It is desired to attain a vacuum of about 10^{-5} mbar in 1 h after exposure of the chamber to atmospheric air. What effective pumping speed *at the entry port* in the base plate must be provided?

The following calculation is valid only for a clean and empty chamber.

† Glass bell jars should have external screens to limit danger following 'implosion'.

Allowance for the outgassing of any work load must be made subsequently. Even so, two assumptions must be made regarding (i) the outgassing rate of the ss inner surface after 1 hour of pumping (virtual leak) and (ii) the real inleak of air from outside, due perhaps to imperfections in the elastomer seals and to leaks in the service entries in the service collar. For (i) we shall take 2×10^{-8} mbar1cm^{-2}s^{-1} after 1 h (Power 1966, p 397) and for (ii) 10^{-5} mbar1s^{-1}. Both these assumptions are guesses. As to (i) the outgassing of ss depends greatly on its history and surface finish and is *not* a 'constant' of the material, while (ii) is a figure which experience shows to be a fairly large leak, and perhaps, therefore, a safe upper limit for reasonably well constructed apparatus.

The volume of the top hat is about 40 l and its internal surface about 0.7 m^2. The virtual leak after 1 h is therefore 14×10^{-5} mbar1s^{-1} and the real leak 1×10^{-5} mbar1s^{-1}, so the assumed value of the real leak will only increase p (the pressure after 1 hour of pumping) by about 7% above the no-real-leak value.

Hence, if S_c is the required pump speed at the work chamber

$$p_1 S_c = \text{throughput in } PV \text{ units per unit time}$$

$$= 15 \times 10^{-5} \text{mbar1s}^{-1} \tag{5.1}$$

and

$$S_c = 15 \, 1 \text{s}^{-1}.$$

Also p_∞, the pressure after infinite time, when the virtual leak has become zero, would be given by $p_\infty \times 15 = 1 \times 10^{-5}$, or $p_\infty = 6.7 \times 10^{-7}$ mbar. Choice of backing pump depends on several factors. In any case, it should retain a pumping capability down to the maximum backing pressure at which the diffusion pump will operate. Assuming that oil diffusion pumps attain their maximum throughput (see § 3.3.4) at an inlet pressure of the order of 10^{-2} mbar, and require a backing pressure of 3×10^{-1} mbar, a rough guide to the speed S_b of the backing pump required to match conditions of maximum throughput during pump-down would be

$$S_c \times 10^{-2} = S_b \times 3 \times 10^{-1}$$

or

$$S_b/S_c = 1/30. \tag{5.2}$$

A considerably smaller ratio of S_b to S_c than that indicated by equation (5.2) is often quite satisfactory if maximum throughput during the early stages of pump-down is not necessary. The intention here is merely to indicate the type of reasoning, the actual figures used depending on the pump-down schedule required. Applying equation (5.2) to the present case gives $S_b = \frac{1}{2} 1 s^{-1}$ or $30\, 1\, min^{-1}$, the latter unit being conventional for backing pumps.

The required speed, given by equation (5.1) is the speed at the entry port of the work chamber, but if a baffle valve is interposed between the work chamber and the diffusion pump the effective speed is usually reduced by about a factor of 2. Allowing another factor of 2 for contingencies, a reasonable estimate of the unbaffled diffusion pump speed for the case of the clean empty chamber would be $60\, 1 s^{-1}$, plus (as explained above) an allowance for the degassing of the chamber contents. It thus appears that precise calculation of the required pump capacity is in general not possible unless experience of the degassing characteristics of the chamber contents is available. However, the type of procedure outlined at least enables an estimate of the *minimum* reasonable pumping capacity, to which must be added an allowance for the outgassing of the contents of the work chamber. In the absence of firsthand data on this, a preliminary guess may be made with the aid of Power (1966, table 5.2, pp 118–9), which itself refers to other sources of data. For instructive examples of pump-down calculations see Steinherz (1963); see also Harris (1977).

5.10 Leak Detection

There are methods for the detection of a leak but not of its location, methods which indicate the location using equipment forming part of the plant, and methods using instruments applied specially to the plant for the detection and location of its leaks. The subject has been reviewed by Turnbull (1965).

5.10.1 Detection of presence, but not location

One procedure consists simply in determining the pressure/time curve of the suspect apparatus when isolated from its pumps. If the pressure rises linearly with time and appears to continue to do so indefinitely, the leak is probably real rather than virtual. If LN$_2$ is applied to a cold finger on the apparatus and most of the gas accumulated (say) overnight disappears, then at any rate most of the leak is virtual and probably consists of desorbed H_2O or another condensable gas not forming a major constituent of the external atmosphere, but derived from the walls and the contents of the work chamber.

Another procedure (McIlraith and Scott 1966) depends on the fact that the speed of a rotary pump is much greater for air than for a condensable vapour.

5.10.2   The bubble method

The apparatus to be tested is immersed in or painted with water, to which soap or some other surface-tension-reducing agent has been added, and internally pressurised by air. A leak, if sufficiently large, causes a bubble to form at its outlet. A brush should be used to clear the submerged surface of bubbles formed on immersion, which would confuse the issue. According to Dennis and Heppell (1968, p 183) the minimum detectable leak is 10^{-4} mbar l s^{-1}.

5.10.3   Use of Tesla discharge

The discharge of a Tesla coil (see § 4.6.2) can locate large leaks (of the order of 10^{-3} mbar l s^{-1}) in glass systems if its high tension probe is passed slowly over suspected parts. The normally bushy discharge jumps to the pinhole leak through which it enters the vacuum system, and in doing so the leak is shown up as a brilliant white spot or line in the glass.

The Tesla discharge can also be used to detect a leak by holding it near the glass part of the apparatus, causing a luminous glow within, while the suspected leaking areas (which may be glass or metal) are painted with a volatile hydrocarbon such as ether. When the leak is reached, the hydrocarbon enters the apparatus and the colour changes from the purple–red characteristics of air to a greyish blue. The virtual leak from the H_2O-laden walls of a newly evacuated apparatus will often give a blue–white discharge which may make difficult the interpretation of a painting-with-hydrocarbon search for a real leak.

5.10.4   Use of instrumentation already part of the apparatus

If the leak or leaks are such as to prevent the attainment of pressures lower than about 10^{-3} mbar, the pressure is in the range of a Pirani gauge. As the response of this gauge depends upon the nature of the gas as well as its pressure see § 4.4.1, leaks may be located by probing the outside of the apparatus with a fine jet of a gas to which the gauge response differs from its response to air, and watching the gauge for a sudden change of reading.

If the ratio between leak rate and pump speed is small enough to cause the equilibrium pressure to be 10^{-4} mbar or less, the hot-cathode ionisation gauge can, without undue risk of filament burn-out, be used to locate leaks in the same way as the Pirani gauge. If He is the search gas, the indication of the gauge drops when the leak is located, but if butane is used, it rises.

If evacuation is via a sputter ion pump, the pump itself may be used as a leak detector, since the ion pump current depends not only on the pressure of the gas in the pump but also on its species. Helium gives the largest increase in current (since it is less well pumped than air) and O_2 the largest decrease. For detailed analysis of this method, see Dennis and Heppell (1968, pp 190–1). Carpenter and Watts (1970) describe a simple method of leak detection, which exploits the difference at LN_2 temperature between the adsorption of a molecular sieve for air and for He.

5.10.5 *Use of special instrumentation*

There are a number of special instruments for leak detection, of which two types may be mentioned here.

(*a*) Types in which there is, between a special gauge and the rest of the vacuum system, a barrier impermeable to air and to all gases likely to be found in the system, but permeable to the probe gas applied to the apparatus from outside. The gauge then indicates when the probe gas enters the system. Examples of this type are the palladium barrier detector (Turnbull *et al* 1962, pp 176 ff) or a cooled charcoal barrier (Kent 1955), both of which are permeable to H_2 but not to most other gases.

(*b*) The mass spectrometer type, in which the probe gas is usually He and the spectrometer (tuned for He) is connected to the apparatus being tested for leaks, a signal being obtained when a He jet impinges on the entrance to the leak. Sensititivity of detection is high (figures of order 10^{-11} mbar l s^{-1} are quoted), but comparison of the sensitivities of various instruments must be made cautiously unless the exact conditions of use are specified. For instance, the time of application of He to produce a given signal is clearly relevant (Turnbull 1965). For a critical discussion of sensitivity of leak detection see Turnbull *et al* (1962, pp 176 ff).

References

Allsopp H J 1961 *Vacuum* **11** 39
Altemose V O and Kacjon A R 1979 *J. Vac. Sci. Technol.* **16** 951–4
Barton R S and Govier R P 1965 *J. Vac. Sci. Technol.* **2** 113–22
Bills B G and Evett A A 1959 *J. Appl. Phys.* **30** 564–7
Carpenter L G and Watts M J 1970 *Vacuum* **20** 197
Dennis N T M and Heppell T A 1968 *Vacuum System Design* (London: Chapman and Hall)
Donaldson E E 1962 *Vacuum* **12** 11–4
Elsey R J 1975 *Vacuum* **25** 299–306, 347–61
Espe W 1966–8 *Materials of High Vacuum Technology* vol 1–3 (Oxford: Pergamon)
Green G W 1968 *The Design and Construction of Small Vacuum* (London: Chapman and Hall) ch 3
Harris N S 1977 *Vacuum Engineering* (Watford: London Caledonian Press) pp 49–50
Hartman R L 1967 *Rev. Sci. Instrum.* **38** 831
Holland L 1961 *The Properties of Glass Surfaces* (London: Chapman and Hall)
Holland L *et al* 1974 *Vacuum Manual* (London: E and F N Spon) pp 2 ff
Kendall B R F and Zabielski M F 1966 *J. Vac. Sci. Technol.* **3** 114–9
Kent T B 1955 *J. Sci. Instrum.* **32** 132–4
Kohl W H 1964 *Vacuum* **14** 333–54
Laurenson L private communication
Lewin G 1965 *Fundaments of Vacuum Science and Technology* (New York: McGraw-Hill) ch 7
McIlraith A H and Scott A D L 1966 *J. Sci. Instrum.* **43** 961
Mog D 1976 *Vacuum* **26** 25–9
Power B D 1966 *High Vacuum Pumping Equipment* (London: Chapman and Hall)
Raj K and Grayson M A 1981 *Vacuum* **31** 151–4
Robinson N W 1968 *The Physical Principles of Ultra-High Vacuum* (London: Chapman and Hall)

Rosebury F 1965 *Handbook of Electron Tube and Vacuum Technique* (London: Addison-Wesley)
Sayers J F 1960 *J. Sci. Instrum.* **37** 203–5
Steinherz H A 1963 *Handbook of High Vacuum Engineering* (New York: Reinhold) ch 11
Turnbull A H 1965 *Vacuum* **15** 3–11
Turnbull A H, Barton R S and Riviere J C 1962 *An Introduction to Vacuum Technique* (London: George Newnes)

6   Ultra-high Vacuum

6.1   General

Unless special precautions are taken, the surface of a solid is never clean but always covered by adsorbed foreign molecules. To illustrate the problem, consider an initially clean solid surface in a pressure of 10^{-9} mbar of gaseous nitrogen at room temperature. Assume that a monolayer is 8.26×10^{14} molecules/cm^2 and that every molecule which hits the clean surface is adsorbed. The bombardment rate is (see equation (AI.7)) 2.7×10^{11} molecules/cm^2/s, hence 5% of the surface is covered in 153 s.

It is evident that a UHV technique is necessary if it is desired to carry out research or perform manufacturing processes on *clean* surfaces.

If the definition of ultra-high vacuum is taken (see § 1.1) as a pressure of 10^{-8} Torr or less, UHV was probably achieved many years ago. However, this was not then recognised, because the soft x-ray effect in ionisation gauges (see § 4.5.1) prevented the measured collector current from falling below a value which, interpreted in terms of pressure, corresponded to about 10^{-7} Torr. However, Nottingham (1947) pointed out that the real *positive ion* current was smaller than the measured collector current, since the latter was augmented by photoelectrons *leaving* the collector. The introduction of the Bayard–Alpert gauge (see § 4.5) allowed measurement down to 10^{-10} mbar, and subsequent developments have extended the lower limit by several decades.

Once it became possible to measure low pressures, effort was directed to methods of obtaining them. Oil or mercury diffusion pumps equipped with LN$_2$-cooled traps had ultimate pressures of 10^{-10} mbar or below, but material outgassing prevented the attainment of UHV in a work chamber in a reasonable pumping time. Hence attention was concentrated on the materials and construction of chambers and demountable seals bakeable to several hundred degrees centigrade, so that the room-temperature outgassing rate after baking might be sufficiently small compared with the available pump speed to enable UHV to be attained reasonably quickly.

Thus UHV technique is largely concerned with the material and construction of work chambers and the measurement of low pressures attained in them. A useful general review of UHV technique, with special reference to pumps and to total and partial pressure gauges, is given by Craig (1967). Carter *et al* (1972) have reviewed the factors limiting the attainment of UHV below 10^{-10} mbar.

6.2 Pumps

The production of UHV can be achieved by several types of pump already described in chapter 3. The subject has been treated in specialised works by Redhead *et al* (1968, ch 11) and by Robinson (1968), so only a few general remarks are made here.

Diffusion pumps with working fluids of mercury or oil can be used with success with well designed cold traps, suitably pre-baked. The advantage of mercury is that, being an element, it cannot decompose as maltreated organic pump fluids will. But a cold trap should preferably be interposed between the backing pump and the mercury pump, as well as between the latter and the work chamber, if certainty of exclusion of organic vapours is desired.

Since, at a given temperature, the vapour pressure of pump oils is much lower than that of mercury, a mechanically refrigerated cold trap at $-40\,°C$ interposed between the diffusion pump and the work chamber allows (see Tolmie 1966) pressures of about 10^{-10} mbar to be attained. The use of mechanical refrigeration, rather than LN_2, has considerable advantage in convenience.

A popular UHV system consists of a sputter ion pump, initially evacuated to its starting pressure by a sorption pump (see § 3.5) and 'topped up' (i.e. pressure further reduced) by a titanium sublimation pump (see § 3.4.2). The latter need only be operated intermittently, as at low pressures a freshly sublimed layer of Ti remains active for a considerable time.

Recent developments in turbo-molecular pumps have made them a convenient method of obtaining UHV. Their rather high capital cost is partly offset by the fact that in most cases they do not require refrigerated vapour traps.

6.3 Methods of Pressure Measurement

6.3.1 The problem

The difficulty of UHV pressure measurement lies in the presence of spurious ion collector currents, due to causes such as the soft x-ray effect, rather than in the small value of the current itself. For example, in hot-cathode gauges with a sensitivity (see § 4.5) of 10 mbar^{-1} and an ionising electron current of 1 mA, the ion current corresponding to 10^{-12} mbar is 10^{-14} A. However (see Redhead *et al* 1968, p 295) currents of 10^{-15} A can be measured with a response time of about 1 s, and lower currents can be measured using secondary electron multiplier techniques.

6.3.2   *Hot-cathode gauges*

The lower pressure limit of the Bayard–Alpert gauge (BAG) in its original form has been extended by Redhead *et al* (1968, p 312) by modulation of the collector current. The essential feature is the introduction, within the grid space and parallel to the ion collector, of a wire whose potential can be changed rapidly from that of the grid (when it collects no positive ions) to that of the collector, when it reduces the collector current by some 30–40%. This change, however, has little effect on the residual component of the collector current due to soft x-rays or other disturbing effects. The percentage modulation can be determined at pressures so high that the residual current is negligible compared with the true ion current. Using this modulation factor, the observed collector current at low pressures can be separated into the true ion current (from which the true pressure can be calculated) and the disturbing residual current.

Redhead and Hobson (1965) have developed two other forms of UHV gauge: the suppressor and the extractor types. In the suppressor gauge, the emission of photoelectrons from the ion collector is prevented by an electric field at its surface, which feature, combined with a modulator electrode, has allowed the measurement of pressures of about 10^{-14} Torr.

In the extractor gauge the collector is a short fine wire. The ions pass through a small hole in a shield and are electrostatically focused on the collector, which is thus rendered partially immune from soft x-ray effects. The desorption of gases by electron bombardment of the grid is much less than in the BAG or suppressor gauge. This is a decided advantage when measuring very low pressures. The lowest pressure measured by the gauge is 9×10^{-13} mbar.

The orbitron gauge has already been mentioned (§ 4.5.2). It has a low x-ray limit and, since its sensitivity is so high, requires only a small electron current, and therefore temperature induced interaction with the filament and desorption from the anode under electron bombardment are minimised.

In the bent-beam gauge developed by Helmer and Hayward (1966), a beam of ions, extracted from a grid cage in which they are produced by electron bombardment, is bent through a 90° arc by an electric field and reaches a collector which cannot 'see' the grid cage. A Cary vibrating reed electrometer measures the ion current to the collector, photoelectric emission from which is prevented by a suppressor grid.

Hot-cathode low-pressure gauges have been developed in which a magnetic field both increases the electron path and suppresses the emission of photoelectrons from the ion collector surface. Details of two gauges of this type, developed by Lafferty and Klopfer, are given by Redhead *et al* (1968, pp 319 ff).

6.3.3 Cold-cathode gauges

Two types, having crossed electric and magnetic fields, are the magnetron and the inverted magnetron UHV gauges. Both have cylindrical symmetry, the electric field being radial and the magnetic field parallel to the axis, and of such magnitude that electrons emitted from the cathode and attracted to the anode are forced by the magnetic field (see § AII.16) to move in curved trajectories and return to the cathode, unless they lose energy by collision with a gas molecule.

In both types, the construction is such that only the ion current and not the field emission current (see § AII.14) is measured. The latter, however, is useful in starting the discharge at low pressures.

In the magnetron gauge, the cold cathode is an axial cylinder having two circular end discs and the anode is a coaxial squat cylinder. With an applied potential of 6 kV and an axial magnetic field of 1 kG the sensitivity is of the order of 10 A mbar^{-1}. In the inverted magnetron (Hobson and Redhead 1958), the anode is a cylinder and the cathode is a coaxial cylinder surrounding it. The applied voltage is about the same as that of the magnetron gauge, but the axial magnetic field is 2 kG.

According to Redhead *et al* (1968, p 334), who have developed both types, the inverted magnetron gauge appears to be more stable and less noisy than the magnetron gauge.

The advantage of cold-cathode gauges for UHV measurement is that, provided the geometry excludes the measurement of the field emission electron current, the electron current is a function of the gas pressure.

Hence, any soft x-ray effect produced by electrons striking the anode is also a function of gas pressure. Thus there is no soft x-ray limit.

6.3.4   *Partial pressure analysis*

When it is desired to know not only the overall nitrogen equivalent pressure (see § 4.1) but also the composition of the gaseous environment, a partial pressure analysis is necessary.

A partial pressure gauge is essentially a mass spectrometer specially adapted to identify the residual gases in a vacuum system. It has a higher sensitivity and is more compact than a conventional mass spectrometer, but the mass range and resolution are relatively limited.

The subject of both total and partial pressure measurements in UHV has recently been reviewed by Weston (1979, 1980) in two articles which should be read by anyone considering working in the field. Weston points out that the composition of residual gas in a UHV system is unlikely to be largely N_2, derived from the atmosphere before evacuation, and therefore *total* pressure measurement may not give meaningful information about the environment.

6.4   Construction and Materials of UHV Systems

Much of the information already given in chapter 5 is applicable to UHV problems. The present section adds a few comments of special relevance.

A principal requirement of materials and couplings for UHV is that they must be bakeable to several hundred degrees centigrade to reduce subsequent outgassing. Except for small non-demountable Pyrex apparatus, this means vacuum envelopes of ss in practice and avoidance within the work chamber of any material with a high vapour pressure. For example, hard solders, if used, must not contain zinc or cadmium.

If the vacuum chamber consists, as it usually does, of a number of stainless steel components bolted together, with metal/glass or metal/ceramic electrical lead-ins, these should be individually tested, before assembly, on a leak detector, preferably of the He-tuned mass spectrometer type. It is good practice to bake out ss parts in an auxiliary vacuum furnace before leak testing, in order to 'open up' any potential leaks, which can be dealt with more easily at this stage.

Leak consideration should include the fact that certain materials, notably Pyrex and fused silica, are permeated by He and also take it into solution. Since the partial pressure of He in the atmosphere is about 5×10^{-3} mbar, its presence may be pressure limiting in an all-Pyrex apparatus.

Redhead *et al* (1968, pp 107 ff) give an instructive calculation of the time variation of He partial pressure in a spherical Pyrex vessel, of volume 1 l and wall thickness 1.5 mm, evacuated by a $0.1 \, \mathrm{l \, s^{-1}}$ pump. The attainment of a linear concentration gradient of He in the Pyrex, and hence of a constant pressure of He in the vessel, would take a long time at room temperature. After 10 h pumping, the pressure would still be about one order greater than the equilibrium, i.e. He would still be emitted from the inner surface more quickly than it entered the outer one.

However, a 10 h bake at 500 °C during which the pressure would rise to a steady value of 8×10^{-9} mbar (owing to the increase of the diffusion coefficient of He with temperature) would establish a linear concentration gradient. Because the *solubility* of He in Pyrex is virtually temperature independent, this would allow the final equilibrium pressure of 11.3×10^{-12} mbar to be reached immediately on return to room temperature.

The frictional force between surfaces in contact is greatly increased in UHV, because the adsorbed layer of gas or other contaminant, which is normally present and acting as a lubricant, is absent or much reduced. Hence bearings and sliding surfaces easily seize, i.e. 'cold welding' takes place.

The effect (see e.g. Winslow and McIntyre 1966, Rowe 1957) can be minimised in several ways. One of these is the use of MoS_2 as a dry lubricant, though for really clean work the possibility that dissociation may produce free S is disturbing. Another expedient is the use of a thin layer of some low-vapour-pressure element which is easily sheared and non-reactive, such as Au (Spalvins and Backley 1966). A third method is the use, without an intervening lubricant, of combinations of two metals found to have a small tendency to seize. Osborn tried a number of bearing combinations of which a pure Fe shaft in a Ag bearing was found to be best; see Carpenter (1965).

References

Carpenter L G 1965 *Proc. Inst. Mech. Eng.* **179** Part 3G 165–71
Carter G, Almour D G and Chernatony L de 1972 *Vacuum* **22** 643–9

Craig R D 1967 *Vacuum* **16** 70–8
Helmer J C and Hayward W H 1966 *Rev. Sci. Instrum.* **37** 1652–4
Hobson J P and Redhead P A 1958 *Can. J. Phys.* **36** 271–88
Redhead P A and Hobson J P 1965 *Br. J. Appl. Phys.* **16** 1555–66
Redhead P A, Hobson J P and Kornelsen E V 1968 *The Physical Basis of Ultra-High Vacuum* (London: Chapman and Hall)
Robinson N W 1968 *The Physical Principles of Ultra-High Vacuum* (London: Chapman and Hall) ch 2
Rowe G W 1957 *Inst. Mech. Eng. Proc. Conf. on Lubrication and Wear* pp 333–8
Spalvins T and Backley D H 1966 *J. Vac. Sci. Technol.* **3** 107–13
Tolmie E D 1966 *J. Sci. Instrum.* **43** 954–7
Weston G F 1979 *Vacuum* **29** 277–91
——1980 *Vacuum* **30** 49–67
Winslow P M and McIntyre D V 1966 *J. Vac. Sci. Technol.* **3** 54–61

7 Applications of Vacuum Technology

7.1 General

Vacuum technology was originally a tool of laboratory research used, for example, to remove the normal atmospheric gases so that the empty space (vacuum) could be filled with other gaseous species on which it was desired to experiment. A typical laboratory use was concerned with the investigation of electrical discharges through gases at low pressures, in which the mean free paths of atoms, ions and electrons were large compared with their values at atmospheric pressures. The discovery of the electron, and the determination of its charge/mass ratio was made possible by this means.

Today, high vacuum has become such a common research tool that it would be difficult to find any sizable research laboratory not possessing a number of pieces of vacuum-producing equipment. Thus, the present uses of vacuum *in research* are too numerous and varied to catalogue and will not be mentioned in this chapter.

However, on account of their economic importance and intrinsic interest, some industrial applications will be briefly described. These applications are mainly of two kinds.

One is concerned with the production, often in large numbers, of articles which are evacuated (e.g. thermionic valves and dewar flasks) or having been evacuated of normal atmospheric air are refilled with another gas or mixture of gases (e.g. in gas-filled filament lamps or discharge lamps).

The other main application is to manufacturing or processing plants, in which vacuum is used to produce articles which in their final form are not evacuated; examples range from vacuum-melted steel ingots (weights of the order of 200 tons) to evaporated electronic micro-circuits. For an excellent treatment of certain applications in more detail than the scale of the present book allows, see Pirani and Yarwood (1961). See also Harris (1977) who lists typical applications of vacuum technology and specifies the pumping systems used.

7.2 High Vacuum as an Electrical Insulator

The physical basis of high-voltage vacuum insulation is treated by Latham (1981).

Vacuum relays, switches and interrupters are commercially available, the advantage of vacuum being that the breakdown voltage for a given gap is greater *in vacuo* than in air, and when breakdown does occur, arcing is diminished and recovery time shortened.

High-voltage circuit breakers for primary distribution systems evacuated to lower than 10^{-4} mbar are described by Fink *et al* (1979).

Variable capacitors, in which the dielectric is a high vacuum may be used as substitutes for air dielectric capacitors (geometrically much larger) because the breakdown voltage at a given electrode spacing is higher *in vacuo* than in air.

In one type manufactured for RF use by the English Electric Valve Co Ltd (to whose courtesy I am indebted for technical information) the electrodes consist of sets of interpenetrating coaxial metal cylinders. The vacuum envelope is partly ceramic or glass and partly bronze bellows, the latter allowing the degree of interpenetration to be varied by axial movement of one set of cylinders (i.e. one electrode). Hence the capacity is variable but vacuum integrity preserved. Manufacture includes degassing by heat during evacuation by diffusion pumps. Getters are not used, since any films of deposited material tend to increase the possibility of field emission (see § AII.14), which in turn leads to premature breakdown. Capacitors are available with peak rated voltages between 3 and 30 kV and the possible range of variable capacity is of the order 10–10^3 pF. Up to 30 MHz the maximum RF currents (RMS) range from 20 A for convection cooled bellows to 500 A for water cooled bellows, the cooling medium being on the atmospheric side. The degree of vacuum is 10^{-7}–10^{-8} mbar and the nominal clearance between adjacent cylinders is 1 mm per 10 kV of peak rated working RF voltage.

Vacuum relays with rated operation voltages up to 100 kV (DC) are commercially available from Walmore Electronics Ltd (GB) with a maximum operating time of 50 ms and a maximum stable contact resistance of 0.015 Ω. The use of vacuum interrupters for switching and circuit breaking has been reviewed by Reece (1968).

7.3 Separation by Vacuum Distillation

Distillation involves the evaporation of molecules from a liquid and their recondensation on a cooled collector. If this process is carried on in air at atmospheric pressure most of the evaporating molecules will collide with air molecules immediately above the liquid surface and return to the liquid phase. To distil *at atmospheric pressure* liquids of molecular weight M > 200 it is necessary to heat them to a temperature such that the molecules are decomposed. Hence, distillation in air at atmospheric pressure is usually confined to molecules of M < 300.

If the ambient pressure is reduced to (say) 10^{-3} mbar, the mean free path (see § AI.9) in the vapour phase becomes about 60 mm and passage from evaporator to condenser is relatively unimpeded. This is one advantage of distillation under low external pressure.

The energy necessary for the liberation of molecules from the liquid to the vapour phase consists of two terms. One is the energy needed to cross the liquid–vapour boundary (i.e. to 'defeat surface tension'), and the other is needed to expand in the gas phase in order to come into equilibrium with the external pressure. If the external pressure is low the latter term is negligible and the temperature necessary for distillation is reduced.

The practical result is that materials can be distilled over a wide range of temperatures without decomposition. All the components of a mixture have some finite rate of evaporation, and separations at low pressure are less sharp than those achieved at higher pressure.

On the other hand, the range of materials which can be distilled is extended to include those with molecular weights in the range 200 to 600. Examples include fat soluble vitamins, steroids, waxes and numerous chemical intermediates (see Watt 1961, 1963). Whereas vacuum distillation can be carried out in conventional apparatus at a reduced working pressure (say 1 mbar), molecular distillation requires specially designed equipment which will allow an unobstructed path from the evaporator to the condenser for the distilling molecules and a pressure of perhaps 10^{-3} mbar. I am indebted to Mr P Ridgway Watt for information on the subject.

7.4 Freeze Drying

When wet solids containing dissolved or suspended substances are dried in the ordinary way by heat, the removal of water often results in translocation of the substances and the process is not reversed on re-wetting. Freeze drying is a method of removing water by sublimation from the ice phase and translocation is avoided. The material is frozen and the ice sublimes under vacuum to leave a porous soluble residue. To improve stability further the substance is stored away from oxygen and moisture at room temperature or under mild refrigeration until required for use, when its properties are substantially restored by the addition of cold water. A characteristic of freeze drying is that a frozen front or interface retreats from a dry outer boundary that retains the contour of the original frozen material. When properly conducted, migration of constituents, shrinkage and consequent insolubility are avoided. Essential requirements are that the material is first cooled to attain a rigid, preferably crystalline state, that heat of sublimation (about $2800 \, \mathrm{J \, g^{-1}}$ of ice) is provided (or the material is able to draw thermal energy from its surroundings), and that a boundary layer of vapour is not allowed to accumulate in the region of the front. The last requirement is most conveniently satisfied by operating *in vacuo*.

In essence a freeze dryer consists of a vacuum chamber into which frozen material is placed, or a manifold to which containers of such material can be attached. The residual air pressure in the system is typically reduced to some 10^{-2}–10^{-3} mbar by a two-stage rotary pump and the evolving vapour is trapped on a refrigerated surface which is effectively a cryopump.

Material for drying assumes one of three forms: (*a*) a frozen liquid in contact with the walls of a container in which most of the heat of sublimation passes through the base of the container and through the frozen material to the interface or freeze drying front; (*b*) a fragment of frozen liquid or wet solid from which sublimation can take place through the whole of the outer boundary so that heat reaches the interface only through the dry shell; (*c*) small fragments or pellets similar in kind to (*b*) but forming a fluidised bed in which much of the heat of sublimation is transmitted through the evolving vapour.

To improve heat transfer and accelerate sublimation in cases (*b*) and (*c*), various ways of increasing the system pressure have been devised; they include the continuous or pulsed admission of gas, throttling the vapour flow to the condenser and switching off the vacuum pump at intervals. The maximum pressure beyond which mass transfer is adversely affected is some 6 mbar. The technique was devised in the 1960s when efforts were made to improve the economics of food freeze drying. However, expansion of this application has not fulfilled expectations and growth has been in the traditional field of pharmaceutical and biological preparations which include vaccines, hormones, blood fractions, enzymes and antibiotics for which the important criteria are sterility and survival of therapeutic or biologically active properties. Fortunately, the method devised for accelerating the drying of foodstuffs is also applicable to frozen liquids in containers where heat transfer between an irregular base of a glass vial and a supporting shelf are improved. By this means the sublimation or primary drying phase of the process may be reduced by 30% or more.

A typical primary and secondary (desorption) cycle for material 1 cm deep (or thick) to be dried at $-25\,°C$ is some 10–15 h. The cycle time varies approximately as the square of the thickness of the material.

Other materials which are freeze dried include human and animal tissue on a small scale for optical and scanning electron microscopy (and on a larger scale for surgery), whole small animals for display in museums, waterlogged wooden and leather artefacts (as those being recovered from the Tudor battleship 'Mary Rose'†), water-damaged books and manuscripts (Flink 1972), finely divided metal powders, xerographic pigments and catalytic materials.

Mr T W G Rowe, to whom I am indebted for information, has reviewed freeze drying in a recent publication (Rowe and Snowman 1978). See also Cotson and Smith (1963), Longmore (1968) and Milne and Petrie (1961).

7.5 Manufacture of Lamps and Thermionic and Semiconductor Devices

One requirement in lamp manufacture is the attainment of an acceptable filament life, necessitating the removal of gases such as H_2O which interact chemically with the W filament and the introduction of inert gases (Ar and N_2) which decrease its sublimation rate. The lamps are alternately evacuated by a rotary pump and flushed with N_2 and finally sealed off when containing a mixture of N_2 and Ar at about one atmosphere pressure. Final traces of reactive gas are removed by applying the normal operating voltage which fires a red phosphorus getter.

In thermionic valve manufacture, it is imperative to avoid small partial pressures of gases which may 'poison' the emission of the oxide-coated

† The 'Mary Rose' sank into silt which excluded oxygen. Both oxidation and attack by 'aerobic' organisms were therefore prevented. Longbows and arrow shafts were recovered and restored by freeze drying to the extent that it is now becoming known how far an arrow could be shot.

filament. Evacuation to about 10^{-5} mbar by diffusion and rotary pump is followed by the firing of a getter which reduces the pressure by about another three decades. The vacuum technique of cathode-ray-tube manufacture is similar. Television tubes are often pumped by diffusion pumps, backed by rotary mechanical pumps. Silicone oils are used in the former since they will stand exposure to the atmosphere when still hot—an important feature for quick cycling. Details of a typical pumping group for final pumping of television tubes are given by Power (1966). Ward and Bunn (1967) give an account of the application of vacuum technique to the manufacture of electron tubes and lamps.

An important example of the commercial exploitation of vacuum is the semiconductor industry, where sputtering, ion implantation and plasma-etching are used on an industrial scale.

The use of vacuum deposited thin-film micro-circuits from the point of view of practical manufacture is reviewed by Gaffee (1965) and their use in silicon chip semiconductor integrated circuits by Glang (1966). See also Brewer (1980) which deals with electron beam technology in micro-electronic fabrication.

7.6   Vacuum Metallurgy

Vacuum technology is of considerable and increasing interest in metallurgy, because many metallurgical processes involve heat, and most metals react with atmospheric air when heated.

Vacuum metallurgical processes can be considered under two main headings—the removal of dissolved gases or other unwanted species from metals, and the carrying out of operations which the presence of air would impede or prevent.

A general review of vacuum processes in engineering is given by Saunders (1980) in which (incidentally) the importance of vacuum in the processing and repair of turbine blades is emphasised.

The reduction of the hydrogen content of steels by vacuum treatment is of increasing importance. Very small concentrations of dissolved hydrogen in high-tensile steel can lead to embrittlement (Rollason 1961) and vacuum degassing in the molten state produces a significant improvement. In 1968, more than 250 plants were in operation, in some of which ingots up to 250 tons could be cast. The treatment is also used for the controlled de-oxidation of steel melts and permits the introduction of reactive alloying elements which would otherwise be partly and uncontrollably lost by oxidation. For details, and other information concerning vacuum melting, see Winkler (1968).

The application of vacuum technology to steel making is reviewed by Barraclough (1972) who gives an interesting account of the historical background and deals with modern industrial plants. Electron beam guns, developing powers of several hundred kilowatts per gun, have important metallurgical uses, especially in the melting, in water-cooled metal crucibles, of refractory metals such as Ta at temperatures so high that reaction with conventional ceramic crucibles would produce unacceptable contamination (Dietrich *et al* 1968). Large-scale electron beam processes in melting steel, degassing, slab conditioning, annealing and heat treatment, and evaporative coating are reviewed by Coad *et al* (1966).

Electron beam welding (a specialised application of electron beam guns) is characterised by extreme power *density*, good controllability, and the possibility of welding in cavities of a large depth-to-width ratio. Although originally developed for small-scale work, where very precisely located welds are required, it is by no means confined to these applications, and it has found ready application to the nuclear, aeronautical and mass production industries where the deep penetration capability has been used for the fabrication of large components and sub-assemblies (such as automatic gearboxes) requiring weld depths of over 25 mm (and beam power of over 8 kW). Weld depths of over 200 mm at 75 kW of beam power are reported, (see Saunders 1982, who deals with electron beam welding in production and gives reasons for its increased use). I am indebted to Torvac Ltd, Cambridge, for technical information.

In the original form of the electron gun, the workpiece to be welded had to be introduced into the vacuum chamber (pressure of order 10^{-4} mbar) in which the electron gun itself operated—a serious limitation on the speed of production and the size of workpiece. This limitation has however (Bakish 1968) been overcome by the introduction of a separate, independently pumped chamber for the electron gun (or by a series of such chambers in the case of atmospheric working) isolated by a valve from the work chamber, which can thus be vented or pumped without disturbing the gun vacuum. In the case of atmospheric working the weldment may be shielded by an inert gas blast (helium being preferred as the electron scattering effect is less marked than with gases of higher atomic number). Electron beam welding of electronic components is discussed by Sanderson (1967).

Vacuum brazing (pressure 10^{-4} mbar) is useful for components small enough to go into a vacuum furnace. It is fluxless, gives good and reliable joints (Harvey and Perry 1963), but is generally higher in capital cost of plant and lower in output than an H_2 furnace of the same charge size. However, it has definite advantages for the brazing of complex structures with interstices, such as heat exchangers, which are difficult to flush adequately with H_2. The advantages, technical and economic, are discussed by Boston (1965).

Vacuum sintering is used to produce metal in a massive form from starting material in which it is in a finely divided state. For example, Ta is obtained as a grey metallic powder by reduction of Ta compounds. This is pressed into bars which are transformed into blanks by sintering or melting in a high vacuum. The blanks are converted into rods, sheets, wires and foils by cold deformation with intermediate high-vacuum anneals.

Vacuum sintering is also used in the fabrication of 'hard metal' tools e.g. WC sintered with cobalt.

The fundamentals of vacuum metallurgy are reviewed by Browne (1971) who gives the thermodynamical background and a list of vacuum metallurgical processes, with a short description of each.

7.7   Coating by Vacuum Evaporation

In some optical instruments, e.g. cameras, the fact that each air/glass surface reflects about 5% of the incident light may be serious. This can be considerably reduced by 'blooming', i.e. by coating the glass surface with a transparent medium of suitable refractive index. Magnesium fluoride or SiO are often used, a thickness of order 10^{-5} cm being applied by vacuum evaporation (Ditchburn 1963) from a silica or Ta crucible.

In micro-electronics (Holland 1965), the passive elements of thin-film integrated circuits (i.e. the resistors and capacitors) are fabricated by vacuum deposition in background pressures of order 10^{-5} mbar or lower. Ni–Cr of thickness of order 10^{-6} cm is used for resistors, and SiO as the dielectric of capacitors; the electrodes and interconnections are often evaporated Al.

Tantalum, which is a high-melting-point metal and difficult to evaporate, has been sputtered (see § AII.18) in pure dry Ar to form resistors and, anodically oxidised, used as the dielectric of capacitors.

Aluminising of the front surface of mirrors is carried out in pressures of about 10^{-5} mbar (Holland and Barker 1965), producing a maximum reflectance for visible light of about 92%. The resistance to tarnish and corrosion is excellent and the mechanical strength reasonable. A layer of SiO is sometimes deposited on the aluminium to give additional protection against abrasion.

Steel is often protected against corrosion by a thin coat of Cd, applied by electrolysis, which also, however, introduces hydrogen, causing loss of strength if the steel is of the high-tensile type (Rollason 1961). Vacuum deposition of Cd carried out under proper conditions of surface cleanness gives adhesion as good as that obtained by electro-deposition and avoids hydrogen embrittlement (Clark and Carpenter 1963).

Plant for the continuous production of steel strip protected on one or both sides by evaporated metal (e.g. Al) has operated successfully on a large scale (Dietrich *et al* 1968). The strip enters the plant at atmospheric pressure and is led, via a number of thin slit apertures through a series of chambers pumped to successively lower pressures, into the evaporation chamber at 10^{-4} mbar and out again to the atmosphere through similar slit aperture systems. It is claimed that this process uses less electric power per unit of material deposited than electrolytic plating.

Vacuum coating, unlike electro-plating, can be applied to electrical insulators. Hence plastics are also coated on an industrial scale by vacuum evaporation (Barker 1963). First, a coating of lacquer is applied to limit the degassing of the plastic base and to produce a smooth surface for the following evaporated metal coating—often aluminium with a thickness of the order of 10^{-5} cm. The final stage is the application of an outer protective coat of lacquer (which may, if desired, be dyed) producing a smooth lustrous decorative finish. The process is used in the automobile, artificial jewelry and toy industries.

However, the vacuum metallising of plastics has special problems arising from degassing which, despite the first lacquer coat, exceeds that of glass by several orders of magnitude. As water or condensable vapours are a major constituent of the outgassing, cryopumping can be usefully employed in conjunction with more conventional equipment.

Holland and Barker (1965) have described one such system in which a 72 inch diameter work chamber 72 inches long is pumped by six combined diffusion/booster pumps (Edwards type 9B4) having a combined speed for air (when baffled) of $9000\,1s^{-1}$, and backed by a $125\,1s^{-1}$ gas-ballasted rotary pump. The cryopump, cooled by LN_2, is in the form of a flat helix of copper tubing having a calculated pump speed for H_2O vapour of about $2 \times 10^5\,1s^{-1}$. The vessel contains a rotating work-holder and, under a plastic load of 1200 acrylic mouldings (each 1 oz in weight), a vacuum suitable for metal finishing (i.e. 6.7×10^{-4} mbar) can be attained in 12 min if the cryopump is used, or in 19 min without it.

7.8 Environmental Testing of Spacecraft

The malfunctioning of a spacecraft in service is extremely wasteful. Hence the necessity for extensive terrestrial environmental testing, *in vacuo*.

The major effects of vacuum on spacecraft include the rapid pressure reduction during launch, the virtual elimination of gaseous thermal conduction in orbit, the volatilisation of materials, or some components of them, and the effects of vacuum on the function of bearings.

The effects of the launch conditions on corona discharge in electrical apparatus and mechanical stresses between the interior of the vehicle and its environment are discussed by Lorenz (1968), who gives details of pumping equipment used to produce a pressure/time profile which reproduces that of the launch.

In orbit, the thermal state of a space vehicle is determined by the balance between the solar radiation received, the radiation it emits to space (effective temperature about 4 K), its internal thermal linkages and its own power dissipation, which is usually small. At pressures below about 10^{-5} mbar, thermal linkage due to gaseous conduction may be neglected (Carpenter 1965), so that as far as thermal testing is concerned this is a sufficiently low internal pressure. Hence the requirements for thermal-balance test chambers are a pressure external to the spacecraft of 10^{-6} mbar and an internal pressure less than 10^{-5} mbar, solar simulation of the order of 1 solar constant† and a heat sink at a temperature such that the radiation from it to the space vehicle is small compared with that emitted by the vehicle itself. Since radiation is proportional to the fourth power of the absolute temperature, the heat sink need not be cooler than 100 K, a temperature conveniently produced by an LN_2-cooled shroud. This also serves as a cryopump for the easily condensed gases and supplements the oil diffusion pumps usually employed.

---

† The solar constant is the amount of solar energy which (in the absence of atmospheric absorption) would fall in unit time on unit area of a surface at the mean distance of the earth from the sun, placed so that the solar radiation falls normally upon it. Its value is approximately 0.13 W cm^{-2}.

Excessive outgassing of materials in space vehicles may make the time taken to attain the necessary vacuum for internal thermal tests (about 10^{-5} mbar) intolerably long. It may also be detrimental if material sublimes from the warmer to the cooler parts, where its presence affects function, or if the loss of volatile constituents degrades the mechanical properties, e.g. loss of flexibility in plastics. While a vacuum of the order of 10^{-5}–10^{-6} mbar is sufficient for thermal testing and to allow sublimation to take place unimpeded (Carpenter 1965), it is not adequate for the testing of bearings and sliding surfaces.

In space, the pressures are in the UHV region (e.g. at 600 km the particle density corresponds to an air pressure at laboratory temperature of about 10^{-10} mbar); in interplanetary space, they are many decades lower. Long exposure may remove the adsorbed layers from surfaces not in pressurised compartments, and increased friction and seizure (see § 6.4) may result. Hence such components should be additionally and individually tested in small UHV chambers, since outgassing of the complete space vehicle makes the attainment of vacua comparable with those in space impracticable in large environmental chambers. Patrick (1981) has discussed the *in vacuo* properties of spacecraft materials.

There are in existence a number of space-simulation chambers in which full-size satellites can be tested. In Great Britain, the largest is that at the Royal Aircraft Establishment, at Farnborough, Hampshire (Earl *et al* 1968). The chamber is large enough to accept any spacecraft which can be included within a 2.5 m diameter sphere, and solar simulation is achieved by six carbon arc lamps shining through six 20 cm diameter ports, capable of giving the equivalent of about $1\frac{1}{2}$ solar constants on a 2.5 m diameter spacecraft. The pumping system consists of two 91.5 cm diameter oil diffusion pumps backed by a rotary pump of capacity $8500 \, \text{l min}^{-1}$. The chamber can be pumped down to below 10^{-6} mbar at which the effective pumping speed is at least $25\,000 \, \text{l s}^{-1}$. An LN$_2$-cooled liner aids the pumping of condensable vapours.

7.9 Vacuum as Thermal Insulation

When the pressure in an evacuated space is reduced to a value such that the average distance of unimpeded motion of the gas molecules is limited by their collisions with *solid* obstructions (as opposed to other gas molecules), the effective thermal conductivity falls below the value given by equation (AI.10). This principle is applied in the domestic picnic flask and on a large scale in industrial containers for liquefied gases, such as LN_2 and LO_2. Typically, the thermal insulation of these large industrial containers consists of an evacuated annular space, filled with a heat-insulating powder, the average size of the interstices of which are small compared with the mean free path appropriate to gas–gas collisions at this pressure (see § AI.9). Thus the condition for reduced thermal conductivity is fulfilled. The powder particles also form a series of radiation shields between the outer vessel at room temperature and the inner one at the temperature of the liquefied gas.

One material of construction is aluminium and the powder-filled space is evacuated via a cold trap (§ 2.6) by a rotary mechanical pump, possibly augmented by a diffusion pump. The service pressure, after sealing off, is 2×10^{-1} mbar when the inner vessel is empty and 10^{-2} mbar when it is filled with liquefied gas, the reduction being due to sorption pumping (see § 3.5).

A modern type of construction for laboratory dewars involves the use of 'superinsulation', consisting of many alternate layers of aluminium foil radiation baffles separated by low conductivity spacer material. See also Snowman (1971) and Tantam (1981). I am indebted to the courtesy of the British Oxygen Company for technical information. For further information *Cryogenic Engineering News* (1969) may be consulted.

References

Bakish R 1968 *Proc. 4th Int. Vacuum Congress* Part 2 (London: The Institute of Physics and The Physical Society) p 464
Barker D W 1963 *Int. Plastics Eng.* **3** 46–52, 116–9
Barraclough K C 1972 *Vacuum* **22** 91–102
Boston M E 1965 *Machinery* **106** 1079–83
Brewer T J ed 1980 *Electron Beam Technology in Micro-Electronic Fabrication* (London: Academic)
Browne R J 1971 *Vacuum* **21** 13–6
Carpenter L G 1965 *Proc. Inst. Mech. Eng.* **179 3G** 165
Clark D and Carpenter L G *GB Patent Specification* 43032/63
Coad B C, Hunt C d'A and Smith R H 1966 *Sheet Metal Industries* **43** 800–7
Cotson S and Smith D B 1963 *Freeze Drying Foodstuffs* (Columbine)
Cryogenic Engineering News 1969 February 18–21
Dietrich W, Gruber H, Sperner F and Stephan H 1968 *Vacuum* **18** 657–63

Dietrich W, Hauff A and Reichelt W 1968 *Proc. 4th Int. Vacuum Congress* Part 2 (London: The Institute of Physics and The Physical Society) pp 573–8

Ditchburn R W 1963 *Light* vol 1 (Glasgow: Blackie) pp 138 ff

Earl A G, Swift R D and Spooner A H 1968 *Spaceflight* **10** 174–82

Fink H, Kohler W and Peche G 1979 *Electrical Rev.* **102** 21–3

Flink J M 1972 *Vacuum* **22** 273

Gaffee D I 1965 *Thin Film Micro-Electronics* (London: Chapman and Hall) ch 6

Glang R 1966 *J. Vac. Sci. Technol.* **3** 37–47

Harris N S 1977 *Vacuum Engineering* (Watford: London Caledonian) ch 6

Harvey J T A and Perry E R 1963 *Welding and Metal Fabrication* February

Holland L ed 1965 *Thin Film Micro-Electronics* (London: Chapman and Hall)

Holland L and Barker D W 1965 *Vacuum* **15** 289–99

Latham R V 1981 *High Voltage Vacuum Insulation* (London: Academic)

Longmore A P 1968 *Proc. 4th Int. Vacuum Congress* Part 1 (London: The Institute of Physics and The Physical Society)

Lorenz A 1968 *Proc. 4th Int. Vacuum Congress* Part 1 (London: The Institute of Physics and The Physical Society) pp 213–8

Milne G R and Petrie D S 1961 *Proc. Symp. User Experience of Large Scale Industrial Vacuum Plant* (Institute of Mechanical Engineers) pp 99–106

Patrick T J 1981 *Vacuum* **31** 351–7

Pirani M and Yarwood J 1961 *Principles of Vacuum Engineering* (London: Chapman and Hall) ch 10–14

Power B D 1966 *High Vacuum Pumping Equipment* (London: Chapman and Hall) p 112

Reece M P 1968 *Proc. 3rd Int. Symp. on Discharges and Electrical Insulation in Vacuum, Paris* 305–14

Rollason E C 1961 *Metallurgy for Engineers* 3rd edn (London: Edward Arnold) p 216

Rowe T W G and Snowman J W 1978 *Edwards Freeze Drying Handbook* (Crawley: Edwards High Vacuum)

Sanderson A 1967 *Electronic Components* October 1135–42

Saunders C 1980 *Vacuum* **30** 167–73

Saunders C G S 1982 *Metal Construction* **14** (**1**) 35–9

Snowman J W 1971 *Vacuum* **21** 27–9

Tantam D 1981 *Cryogenics* December 691–5
Ward L and Bunn J P 1967 *Introduction to the Theory and Practice of High Vacuum Technology* (London: Butterworth) pp 181 ff
Watt P R 1961 *Proc. Symp. on User Experience of Large Scale Industrial Vacuum Plant* (Institute of Mechanical Engineers) p 107
——1963 *Molecular Stills* (London: Chapman and Hall)
Winkler O 1968 *Proc. Fourth Int. Vac. Congr.* Part 2 (London: The Institute of Physics and The Physical Society) pp 454 ff

Appendix I. Some Relevant Formulae in the Kinetic Theory of Gases

AI.1 The Maxwellian Distribution of Speeds

If $f(c)dc$ is the fraction of molecules having speeds between c and $c + dc$,

$$f(c) = \frac{4}{\pi^{1/2}} \left(\frac{m}{2kT} \right)^{3/2} c^2 \, e^{-mc^2/2kT} \qquad \text{(AI.1)}$$

where m is the mass of the molecule, T the absolute temperature, c the speed and k is Boltzmann's constant ($1.38 \times 10^{-23}\,\mathrm{J\,K^{-1}}$). From equation (AI.1) one can deduce $\bar{c}$ (equation (AI.3) below) and also $\bar{c}^2$ (equation (AI.2)), and hence the average kinetic energy of the whole assembly of molecules (equation (AI.4)). When $f(c)$ is plotted against c, it has a maximum at a speed called *the most probable speed* C (which is $(2kT/m)^{1/2}$, and hence 88% of $\bar{c}$) and falls to negligible values at very high and very low speeds. An idea of the *breadth* of the distribution is given by the statement that 87% of the molecules have speeds between twice and half C.

AI.2 The Mean-square Speed

$$\bar{c}^2 = \int_0^\infty f(c)c^2 \, dc$$

$$= 3kT/m. \qquad \text{(AI.2)}$$

AI.3 The Mean Speed

$$\bar{c} = \int_0^\infty f(c) \, dc$$

$$= (8kT/\pi m)^{1/2}. \qquad \text{(AI.3)}$$

AI.4   The Average Kinetic Energy

$$KE = \tfrac{1}{2}m\bar{c}^2$$

$$= \tfrac{3}{2}kT \quad \text{[from equation (AI.2)]}. \tag{AI.4}$$

AI.5   Equation of Perfect Gases

$$P = \tfrac{1}{3}nm\bar{c}^2$$

where P is the pressure and n the number of molecules/cm^3

$$P = nkT \quad \text{[from equation (AI.4)]}. \tag{AI.5}$$

AI.6   The Average Rate of Bombardment of a Surface

$$\beta = \tfrac{1}{4}n\bar{c} \quad \text{molecules/cm}^2/\text{s}. \tag{AI.6}$$

Substitution from equations (AI.3) and (AI.5) for $\bar{c}$ and n gives

$$\beta = P(2\pi mkT)^{-1/2}$$

$$= 3.01 \times 10^{20} \text{ molecules/cm}^2/\text{s for N}_2 \text{ at } 0\,°\text{C and 1 mbar.} \tag{AI.7}$$

AI.7   Quantities of Gas

From equation (AI.5)

$$PV = nVkT$$

$$= \text{total number of molecules in } V \times kT \tag{AI.8}$$

where V is the volume considered. Hence, at a standard temperature—say 25°C (298 K) PV is a measure of the total number of molecules in the volume V. So, quantities of gas are quoted in units of pressure × volume (commonly called 'PV units'), e.g. in mbar l. 1 mbar l contains 2.43×10^{19} molecules at 25°C.

AI.8   Flow Rates

These are often expressed in units of pressure × volume per unit time, e.g. the specific outgassing rate of a surface per unit area is given in $\text{mbar}\,\text{l}\,\text{s}^{-1}\,\text{cm}^{-2}$. It is tacitly assumed, unless otherwise stated, that when quantities of gas are expressed in PV units, the temperature is that of the laboratory, say 298 K or thereabouts.

Leak rates are often expressed in $\text{mbar}\,\text{l}\,\text{s}^{-1}$. $1\,\text{mbar}\,\text{l}\,\text{s}^{-1}$ of a perfect gas at 25 °C corresponds to 2.43×10^{19} molecules s^{-1}.

AI.9 Mean Free Path

The mean free path λ is the average distance travelled by a gas molecule before colliding *with another gas molecule*, and is approximately given by

$$\lambda = \frac{1}{2^{1/2} n \pi \sigma^2} \qquad (AI.9)$$

where σ is the molecular diameter, which does not vary greatly between gases, being of order $10^{-8}\,\text{cm}$, unless the molecule is a large organic one.

Since $P = nkT$ (equation (AI.5)), at a given temperature λ varies as P^{-1} and (at room temperature) a useful rough rule is that λ is inversely proportional to P, being about 5 cm at $10^{-3}\,\text{mbar}$.

AI.10 The Thermal Conductivity K of a Gas

$$K \approx \tfrac{1}{3} m \lambda n \bar{c} C_v \qquad (AI.10)$$

where C_v is the specific heat at constant volume. Hence, at constant temperature ($\bar{c}$ constant), K is independent of n (i.e. of P) as long as $n\lambda$ is constant. When, however, P is reduced to the value where λ is greater than the vessel dimensions and the average free path no longer increases as n is diminished, the effective thermal conductivity decreases with P. Hence, for example, the thermal insulation of the evacuated annular space in a Dewar flask.

Appendix II. Notes on Topics in Physics Relevant to Vacuum Technology

AII.1   Atomic and Molecular Magnitudes

Atoms consist of nuclei of diameter about 10^{-12} cm, surrounded by electron clouds of diameter about 3×10^{-8} cm. So molecules (except large organic ones) can be considered as having a target area of 10^{-15} cm^2 and a surface *monolayer* consists of about 10^{15} molecules cm^{-2}.

In a gas at NTP, the average spacing between the centres of molecules is about 10 diameters as opposed to about one diameter (i.e. 'touching') as in solids and liquids.

AII.2   Units and Constants

$$\text{Charge of electron} = 1.6 \times 10^{-19}\,\text{C}$$
$$1\,\text{eV} = \text{a unit of energy}$$
$$= 1.6 \times 10^{-19}\,\text{J}.$$

A *mole* is an assembly of atoms or molecules whose mass is equal to the atomic or molecular weight of the substance in question, e.g. 1 mole of H weighs 1 g, 1 mole of H_2 weighs 2 g, one mole of O_2 weighs 32 g.

AII.3   Evaporation and Vapour Pressure

Consider a substance placed in an evacuated isothermal vessel (temperature T K) with walls which do not react with it. Under the influence of thermal agitation at temperature T, molecules will be emitted by the substance until a pressure P is built up in the vessel, such that the number of molecules striking and sticking to the substance per unit time per unit area is equal to the number emitted from it. P is called the *saturated vapour pressure* of the substance at the temperature T.

P is a strongly increasing function of T. For example, the vapour pressure of H_2O is 23.3 mbar at 20°C and 1 bar at 100°C. If the conditions are such that the number of molecules returning and sticking to the substance is less than the number emitted, it is said to evaporate or sublime, the former term applying to a liquid and the latter to a solid. However, there is no *essential* difference between the two processes.

The great temperature sensitivity of sublimation is exemplified, for example, by Zn. At 300 K, the rate of recession of the surface would be about one monolayer ($\sim 10^{-8}$ cm) per year, the rate at 600 K being about 4 m per year if sublimation were unimpeded.

AII.4   Outgassing

The emission into the gas phase of a foreign substance from the surface of a solid is called outgassing or desorption. An important example is the emission of H_2O vapour from the surface of glass or metal under vacuum conditions. This, like sublimation, is very temperature dependent, hence the common practice of heating a vacuum vessel, so that its rate of outgassing is much decreased *when it cools down again*. A vessel whose walls or contents are emitting gas is often said to have a *virtual leak*. The term 'outgassing' describes spontaneous emission of gas, while 'degassing' describes its deliberate removal by raising the temperature.

AII.5   Adsorption

This is the reverse of outgassing, being the process of capture by mutual intermolecular attractive force, on impact of gas molecules with adsorbing surfaces. The average probability that an impact will result in capture is called the 'sticking coefficient'. The degree of adsorption coverage under a given set of conditions is determined by the competition between the attractive forces (favouring adsorption), and the thermal motion of the molecules of the solid, tending to cause sublimation.

The amount of adsorption is often specified as the number of monolayers, fractional or integral, of the adsorbed vapour which adheres to the solid substrate.

The adsorbed material is referred to as the *adsorbate* and the substrate as the *adsorbent*.

When the number of adsorbed molecular layers exceeds (say) three or four, the adsorbed layer is indistinguishable from the adsorbate *in bulk*, and may be called a macroscopic film in equilibrium with the normal vapour pressure at that temperature. When the film is a monolayer or less, the vapour pressure with which it is in equilibrium is less than that of the adsorbate in bulk.

Since, in adsorption, the mutual attractive forces between adsorbate and adsorbent do work on the adsorbate molecule as it approaches, there is a release of energy as the adsorbate 'sits down' on the adsorbent. This energy is called the *heat of adsorption*, and varies for different adsorption processes of interest in vacuum technology from about $\frac{1}{2}$ to $\frac{1}{20}$ eV (see § AII.2) per molecule. There are 6.02×10^{23} molecules in a mole, and heats of adsorption are often expressed in kilojoules per mole, 1 eV per molecule being equivalent to 96 kJ mole^{-1}.

AII.6 Thermal Accommodation Coefficient

When gas molecules at temperature T_g hit and rebound from a hot solid at temperature T_s they acquire an amount of energy which depends on mutual intermolecular forces during impact and on the temperatures T_g and T_s. If the average energy, after impact and rebound of the impinging molecules can be characterised by a temperature T'_g, the accommodation coefficient α is defined by

$$\alpha = \frac{T'_g - T_g}{T_s - T_g} \tag{AII.1}$$

i.e. by the ratio of the energy per collision lost by the hot solid to that which would have been lost if the impinging molecule had attained the full temperature of the solid.

Efficiency of energy transfer depends on the mutual reaction between the impinging and *surface* molecules of the solid and hence is influenced by adsorption of foreign molecules on the solid surface. For many adsorption-laden surfaces $\alpha \approx 1$, but for *clean* surfaces it is usually much less. For example, for He impinging on *clean* W, $\alpha \approx 10^{-2}$ (Carpenter and Watts 1975). Hence the calibration for He by a Pirani gauge depends on the surface cleanness of its W filament.

AII.7 Gettering

This is a term used to describe the *irreversible* adsorption of a gas by a solid with which it combines to form a *chemical* bond; e.g. O_2 and N_2 are gettered by hot Ti.

AII.8 Diffusion

Diffusion is a process in which particles (atoms or molecules) move through a solid, liquid or gas by means of a *random walk* mechanism, executed by virtue of their thermal motion.

In gas at pressures at which intermolecular collisions are much more frequent than collisions with the walls of the vessel, the length of each random excursion is controlled by collision with neighbouring particles, and the average length of each excursion is called the *mean free path* (see § AI.9).

In vacua, in which the mean free path is limited by collisions with the walls of the vessel, rather than by collisions with other molecules, the diffusion coefficient, as usually defined in the kinetic theory of gases, is not of much interest. However, the diffusion coefficient of gases *in solids* IS relevant to vacuum technology, since here the mean free path is fixed by collision with particles of the solid, and is of the order of 10^{-8} cm—i.e. the spacing of atoms in solids (see § AII.1). The diffusion coefficient D can be defined in a number of ways which are equivalent. A useful definition for our present purpose is as follows. Consider a group of particles diffusing down the z axis, and all starting from $z = 0$ at $t = 0$. Then, at time t, the mean square distance of the particles from the origin is given by

$$\bar{z}^2 = 2Dt \qquad\qquad (AII.2)$$

D, like vaporisation rates, is a strong function of T. For example, D for H_2 in a certain stainless steel is 5.0×10^{-14} cm^2 s^{-1} at room temperature and 3.5×10^8 cm^2 s^{-1} at 300°C. From the above figures, it may be estimated that the time for H_2 to diffuse 1 mm through stainless steel at 300°C is about 40 h, which gives an indication of the time required to degas a piece of stainless steel (the gas content of which is often mainly H_2) of this thickness. Exact calculation is complicated (see, for example, Lewin 1965).

AII.9   Photons

In some situations, the energy of electromagnetic radiation (e.g. light or x-rays) behaves as if it consisted of discrete packets of radiation, whose energy E is given by

$$E = h\nu$$

where ν is the frequency, h is Planck's constant $= 6.6 \times 10^{-34}$ J s.

AII.10   Electronic Structure of Atoms

An atom consists of a nucleus surrounded by a cloud of electrons (see § AII.1). The nucleus is positively charged and the electrons negatively, the number of electrons being such that the sum of their charges (see § AII.2 above) is equal and opposite to the charge on the nucleus, so that the atom as a whole is, in its normal state, electrically neutral.

The electrons are not all similarly situated with respect to the nucleus; they have different orbits, and those which are nearer to the nucleus are more strongly attracted to it and so are more difficult to remove.

The removal of an electron from an atom is called *ionisation*.

AII.11   Ionisation Potential and Efficiency of Ionisation

The minimum energy (in electron volts) required to remove the most loosely bound electron from a gas atom or molecule is called the *first ionisation potential* of that atom or molecule. The removal may be effected by electron impact, radiation (by a photon having an energy at least equal to the ionisation potential), by temperature, or by a variety of other processes, provided they have the necessary minimum energy.

In vacuum technology, ionisation by electron bombardment is of particular importance, being the basis of the measurement of pressure by ionisation gauges.

The probability that electron impact will result in ionisation depends on the energy of the bombarding electron. If the energy of the impinging electron is less than the first ionisation potential, the probability is zero; it increases with energy and reaches a maximum for most gases in the region of 100–200 eV, thereafter decreasing slightly with voltage. Hence the grid of a hot-cathode ionisation gauge is usually between 100 and 200 V positive with respect to the filament, see § 4.5.1.

AII.12 Thermionic Emission

The evaporation of electrons from hot solid bodies is known as thermionic emission. Like vaporisation and outgassing it is a strongly increasing function of temperature.

It is of importance as it provides the current of bombarding electrons in a hot-cathode ionisation gauge, or for example in an electron beam welder.

AII.13 Photoelectric Emission

Photoelectric emission is the emission of electrons from solids, liquids and gases under bombardment by photons, i.e. electromagnetic radiation. In gases, the minimum energy of a photon necessary to cause ionisation of the gas is given by

$$h\nu = eV_i$$

where V_i is the ionisation potential. The frequency of the minimum photon energy necessary to cause electron emission from a solid is given by

$$h\nu = eV_w$$

where V_w is the work function of the electrons in the metal, i.e. eV_w is the minimum energy which an electron requires in order to escape from the metal.

AII.14 Field Emission

Sometimes called *cold emission*, field emission is the emission of electrons from a metal at room temperature under the influence of an electric field normal to the surface of order $10^7 \, \text{V cm}^{-1}$.

AII.15 X-rays

When electrons bombard a solid, a small fraction of the kinetic energy of the impinging electrons is converted into photons, which are emitted from the solid and are termed x-rays.

X-rays of relatively low frequency, corresponding to energies of (say) $E = h\nu = 100\,\text{eV}$, are called *soft*. They cannot penetrate glass, or air at atmospheric pressure, but have sufficient energy to cause the emission of photoelectrons from metals on which they fall, since the work function V_w of metal is only a few volts (see e.g. § 4.5.1).

Hard x-rays are generated by the impingement of electrons of many kV energy. They are penetrating, and physiological protection requires the use of shielding, e.g. electron guns for welding and metal melting use electrons of energy in the 50–100 keV range and shielding against the consequent x-rays is necessary.

AII.16   The Gyromagnetic or Cyclotron Frequency

If a particle, charge e and mass m enters a magnetic field H having a component of velocity $v_\perp$ perpendicular to H, it is acted upon by a force orthogonal to both $v_\perp$ and H and is therefore constrained to move in a circular arc, radius R, round an axis parallel to H, with an angular velocity given by

$$\omega = \frac{e}{m}H. \qquad (\text{AII.6})$$

The corresponding frequency, $\omega/2\pi$ is called the gyromagnetic or cyclotron frequency, the radius R of the orbit being proportional to the value of $v_\perp$ which gave rise to it.

If, on entering the field H, the particle has in addition to $v_\perp$ a component of velocity $v_\parallel$, parallel to H, the path will be a helix of constant pitch. If however there is an axial electric field E, parallel to H, the pitch of the helix will increase as the particle moves down the electric field and gains energy from it.

AII.17   Cold-cathode Discharge

This may best be described with reference to a specific example. If an increasing DC voltage is applied between two plane electrodes spaced (say) 10 cm apart, the pressure being about 10^{-1} mbar of air, a luminous glow discharge starts at about 300 V. Its appearance is not uniform throughout its length, being different at the cathode and anode ends.

If the pressure is gradually decreased, the current decreases and the voltage required to maintain the discharge rises. Finally the discharge ceases, the current falling to zero and the voltage rising to its open-circuit value.

The mechanism of the discharge is as follows. The discharge is initiated by a few chance electrons, produced e.g. by cosmic radiation or by the natural radioactivity of the surroundings. Under the influence of the electric field between the cathode and the anode, the initial electrons multiply themselves via ionisation by collision with molecules of the gas, and the positive ions so produced impinge on the cathode to produce more electrons, which maintain the discharge. The total current measured in the external circuit is the sum of the positive ion current reaching the cathode and the electron current leaving it and is a function of the gas pressure. The above process, modified by the presence of a longitudinal magnetic field (which lengthens electron paths by causing spiralling, see § AII.16) is the basis of the Penning cold-cathode ionisation gauge and of the sputter ion pump (§§ 4.6.1 and 3.4.1).

AII.18 Sputtering

This is said to occur when, in a gas discharge, the impingement of positive gas ions on the cathode causes uncharged particles of the cathode material to be emitted.

The phenomenon is exploited in sputter ion pumps to produce from the cathode a continuous supply of sputtered Ti atoms. These settle on the walls of the pump and 'getter' the gas. Sputtering is used to deposit films of high-melting-point low-vapour-pressure materials which are difficult to evaporate, e.g. Ta.

AII.19 The Boltzmann Factor

In an assembly of molecules in thermal equilibrium at temperature T, the expression for the fraction that possesses, by virtue of thermal energy, an energy q is dominated by the factor $\exp(-q/kT)$ where k is Boltzmann's constant. The quantity $-q/kT$ is a measure of the ratio of q to the *average* kinetic energy of a molecule of the assembly. If $q \gg kT$, it will (since it occurs in the exponent of the exponential) cause the Boltzmann factor to increase rapidly with T.

Equation (AI.1) exemplifies this. There, $mc^2/2kT$ is a measure of the ratio of the kinetic energy of those molecules having a speed c to the *average* translational kinetic energy of the whole assembly, and the temperature dependence of the speed (and therefore the kinetic energy distribution) is dominated by the term $\exp(-mc^2/2kT)$.

References

Carpenter L G and Watts M J 1975 *J. Phys. D: Appl. Phys.* **8** 1660–73
Lewin G 1965 *Fundamentals of Vacuum Science and Technology* (New York: McGraw-Hill) pp 23 ff

Appendix III. Bibliographical Note

Two important English language journals dealing with vacuum science and technology are:

The Journal of Vacuum Science and Technology, published by the American Institute of Physics
Vacuum, published by the Pergamon Press Ltd, England.

Modern books on Vacuum Technology include:

Vacuum Manual 1974 ed L Holland, W Steckelmacher and J Yarwood (London: E and F N Spon)
Vacuum Technology 1976 A Roth (Amsterdam: North-Holland)
Vacuum Engineering 1977 N S Harris (Watford: London Caledonian)
User's Guide to Vacuum Technology 1980 J F O'Hanlon (New York: Wiley).

Appendix IV. Some Chemical Elements used in Vacuum Technology

Name	Atomic weight	Chemical symbol	Temperature (K) at which vapour pressure is 10^{-3} Torr	Melting temperature (K)	Boiling temperature (K)
Hydrogen	1.008	H	6.0	13.9	20.3
Helium	4.003	He	10^{-1} Torr at 0.98 K		4.22
Carbon	12.011	C	2520		4170
Nitrogen	14.007	N	34	63	77
Oxygen	16.00	O	40	54	90
Neon	20.18	Ne	11	24.5	27
Aluminium	26.98	Al	1355	933	2720
Argon	39.95	Ar	39	83.7	87.2
Titanium	47.90	Ti	1831	1953	3600
Chromium	51.97	Cr	1540	2176	2900
Iron	55.85	Fe	1595	1812	3100
Nickel	58.71	Ni	1630	1726	3100
Copper	63.54	Cu	1415	1356	2850
Zinc	65.37	Zn	563	693	1180
Molybdenum	95.94	Mo	2650	2893	4850
Silver	107.87	Ag	1195	1234	2450
Cadmium	112.40	Cd	494	594	1040
Indium	114.82	In	1100	430	2300
Tin	118.69	Sn	1365	505	2900
Barium	137.34	Ba	810	983	1900
Tantalum	180.95	Ta	3080	3270	5700
Tungsten	183.85	W	3280	3650	5750
Platinum	195.09	Pt	2180	2040	4100
Rhenium	186.2	Re	3060	3450	5900
Iridium	192.2	Ir	2380	2720	4700
Gold	196.97	Au	1470	1336	2950
Mercury	200.59	Hg	289	234	630

Notes

(1) Temperatures are quoted in Kelvins. To convert to degrees Centigrade subtract 273°.

(2) The quoted boiling temperatures are, for most high-melting-point elements, obtained by extrapolating lower temperature experimental vapour pressures to the temperature at which they attain the value of 760 Torr. They are, therefore, subject to uncertainty and merely give an indication of the increase of volatility with temperature.

(3) Vapour pressure data for the more common elements and some common gases over a wider range of temperature are given by R E Honig (1957 *RCA Review* **18** 195–204 and 1960 *ibid* **21** 360–8).
Vapour pressure data for the solid and liquid elements are given by R E Honig and D A Kramer (1969 *RCA Review* **30** 285–305).

(4) The evaporation rate (E) *in vacuo* is related to the vapour pressure P of an element of atomic weight M, if it evaporates as single atoms, by the relation

$$E = 5.8 \times 10^{-2} SP \, (M/T)^{1/2} \, \mathrm{g \, cm^{-2} \, s^{-1}}$$

where T is the temperature in Kelvins and S is the probability that an atom of the element, impinging on its own solid at temperature T, will stick to it. For most metals, S is of the order of unity, but may be considerably less for elements such as carbon, whose vapour is partly polyatomic.

(5) A useful table of the physical properties of the elements is given by A H Turnbull, R S Barton and J C Riviere in their book *An Introduction to Vacuum Technique* (1962, (London: George Newnes)). They also give a similar table (which includes information on weldability) for stainless steels and other alloys used in vacuum work.

(6) 1 Torr = 1.33 mbar.

Subject Index